THE ENVIRONMENT OF THE BRITISH ISLES

AN ATLAS

The Environment of the British Isles

An Atlas

A. S. Goudie and D. Brunsden

CLARENDON PRESS . OXFORD

Oxford University Press, Great Clarendon Street, Oxford OX2 6DP

Oxford New York
Athens Auckland Bangkok Bogota Bombay
Buenos Aires Calcutta Cape Town Dar es Salaam
Delhi Florence Hong Kong Istanbul Karachi
Kuala Lumpur Madras Madrid Melbourne
Mexico City Nairobi Paris Singapore
Taipei Tokyo Toronto
and associated companies in
Berlin Ibadan

Oxford is a trade mark of Oxford University Press

Published in the United States by
Oxford University Press Inc., New York

© *A. S. Goudie and D. Brunsden 1994*

Hardback and paperback editions first published 1994
Paperback edition reprinted 1996

British Library Cataloguing in Publication Data
Data available

Library of Congress Cataloging in Publication Data
Goudie, Andrew.
The environment of the British Isles, an atlas / A.S., Goudie and
D. Brunsden.
"Published in the United States by Oxford University Press Inc.
New York"—CIP t.p. verso.
Includes bibliographical references.
Contents: Geological background—Geomorphology—Climate—
Hydrology—Biogeography and soils—Oceans—The human impact.
1. British Isles—Enviromental conditions—Maps.   2. Physical
geography—British Isles—Maps.   3. Man—Influence on nature—
British Isles—Maps.   I. Brunsden, Denys.   II. Title.
G1812.21.C1G62 1994   (G&M)   333.7'0941'022—dc20   94–20386
ISBN 0–19–874172–3
ISBN 0–19–874173–1 (Pbk)

Printed in Great Britain
on acid-free paper by
Bath Press Bath

Preface

Maps are the greatest triumph of the geographer's art. They portray massive amounts of information in a visible, and therefore highly accessible form.

In this concise volume we have aimed to bring together a series of maps, most of them previously published but available only in a highly dispersed fashion. Without being fully comprehensive—an impossible task given the limited span within which we are working—we hope they give a broad picture of the patterns of phenomena which make up the Environment of the British Isles. We hope that the patterns they display will stimulate enquiry and therefore learning.

These maps are based on the labour of many workers, and where possible we acknowledge the individuals or institutions concerned. We are grateful to them all.

We dedicate the book to The Geographical Association on its Centenary, for this body has done more than any other to ensure the place of Geography in British education. The Association began as a means by which visual, geographical educational aids could be developed and exchanged between school teachers. Indeed, its first publication was a catalogue of lantern slides most of which, still preserved in the archives, were of maps of the British Isles. We are proud, one hundred years later, to continue that service.

A.S.G.
D.B.

Oxford and London

Map production acknowledgements

The drawing of these maps has been the result of the combined efforts of three institutions. In the Oxford School of Geography some maps were drawn by Peter Hayward and Ailsa Allen. Peter also co-ordinated the work of the other two departments involved. At King's College in London some of the maps were drawn by Roma Beaumont and Gordon Reynall.

However, we are especially grateful to David Cooper, Principal Lecturer in Mapping Science at the Luton College of Higher Education for allowing us to benefit from the efforts of the Luton College of Higher Education HND Land Administration (Geographical Techniques) course students whose names follow: Jon Birchall, Robert Helm, Lynda Branham, David Jackson, Jonathan Brown, Patrick Kell, Richard Burnill, Richard Knowles, Kim Bushby, Victoria Larwood, Neil Calvert, Kirsty Lyons, John Chastell, Christopher Machin, Neil Cook, Nicholas Mann, Jacqueline Cope, Nicola McNeil, Sarah Curtis, Keith Moore, Sophie Dawson, Brian Perratt, Peter Elliott, Martyn Potter, Alan Fairhurst, Giles Richards, Mark Farington, Christopher Rimmer, Robert Flather, Caroline Smith, Roger Gale, Lucy Stones, Ranth Ginever, Roger Swaine, Simon Goslin, Glen Swindlehurst, John Gulliver, David Waddington: co-ordinated by Janette Archer.

The maps in this volume are derived from a large number of different sources. Because this work is based on these sources there is some inevitable variability in level of detail and geographical coverage. In some cases the original sources did not present information on certain parts of the British Isles and our maps reflect this fact.

Contents

Geological Background

The tectonic setting

The continental crust of the Earth can be divided into tectonic provinces each with its own geological characteristics and history (Figure 1). The story of the British Isles is very long and complex. The islands occur at the junction of several great orogenic belts of Pre-Cambrian (probably 1300–900 MA*), Caledonian (500–400 MA), and Hercynian (300 MA) age, all of which have been affected by the latest Alpine movements (Figure 2).

This location at the 'geographical crossroads' means that the rocks record a huge and varied span of earth history. The physical geography can therefore only be understood in these terms. To the north-west very ancient rocks from the Lewisian gneiss of Sutherland and the Isle of Lewis are structurally part of the Greenland–Canadian orogenic belt of 2500–1700 MA age. Possibly underlying south-east Britain to form a basement are Archaean rocks of gneissic type. These rocks form the nuclei of the continent upon which all the rest has been built.

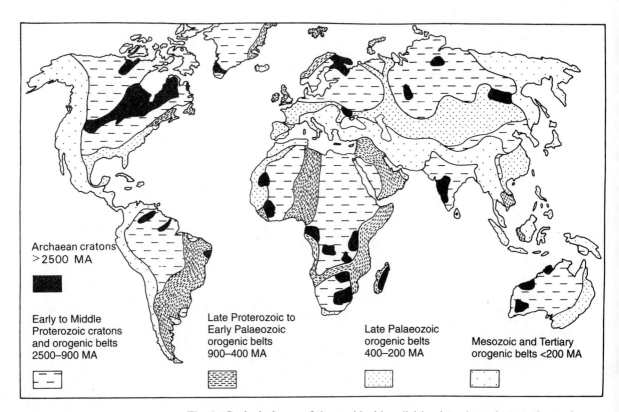

Archaean cratons
>2500 MA

Early to Middle
Proterozoic cratons
and orogenic belts
2500–900 MA

Late Proterozoic to
Early Palaeozoic
orogenic belts
900–400 MA

Late Palaeozoic
orogenic belts
400–200 MA

Mesozoic and Tertiary
orogenic belts <200 MA

Fig. 1. Geological map of the world with a division into the main tectonic provinces. (*Source*: Anderton *et al.* 1979, fig. 1.1.)

* Million years ago.

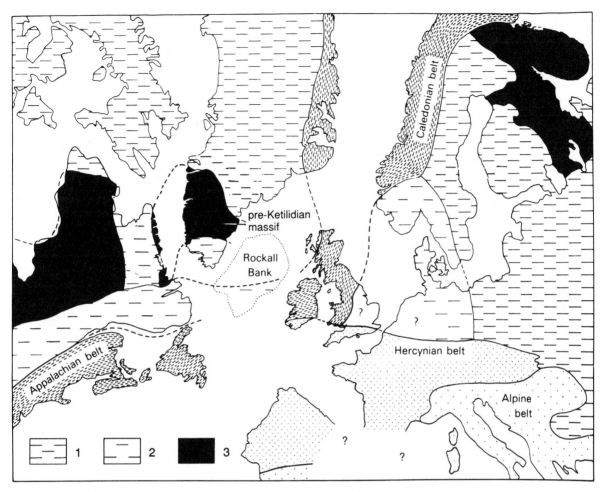

Fig. 2. Geological map of the North Atlantic. Although outlines of present-day coast-lines are shown, the map is actually one based on a configuration before the Atlantic was created. 1–2: early to middle Proterozoic cratons and orogenic belts 2500–900 MA; 3: Archaean cratons >2500 MA.

(*Source*: Anderton *et al.* 1979, fig. 1.2.)

Structure

The creation of the rock skeleton of the British Isles is an incredibly rich and complicated story. Here it is presented in the briefest of general outlines. The description concentrates on the great events because these have determined the main structural outlines and rock types. The development of the structural fabric has also been a remorseless story of great forces, the creation of oceans and mountain chains, denudation, the development of basins, subsidence, uplift, and the establishment of a huge network of faults. Each event has imprinted a legacy which has been reworked time and again to create the present form of Britain. The basis for understanding this complexity requires consideration of larger-scale controls. First, Britain lies at the crossroads of several great tectonic provinces (Figures 1 and 2). Second, it therefore lies at the margin of important geological events.

The early Atlantic 'Iapetus' ocean (Figure 3) opened and closed to produce the Caledonian Mountains and the NE–SW structural trends of Scotland and Ireland (500–400) MA (Figure 4). The great faults of Scotland were initiated at this time. A huge marginal ocean basin, known as the Mid-European Ocean, opened and closed across an area from eastern USA to Bohemia (300 MA). The closure created the E–W trend of the Hercynian mountains, the uplands of south-west England, and the block structures of the main mass of Britain that now dominate the landscape. The NW–SE faults, like the Sticklepath–Lustleigh fault of the West Country, became a dominant feature of the basement at this time.

The next major event (180 MA) was the opening of the Atlantic, which created great tensional forces and the 'stretching' of the region. Huge N–S structures began to develop as the North European craton suffered crustal upwarp and rifting over a very long period of time and the North Sea Basin became a dominant feature which controls sedimentation and river development up to the present day. The overall structure of the British Isles was probably one of a horst-and-graben system with its axis along the Irish Sea and comparable, though less complete, to the Rhine–North Sea Rifts. Seen in this way the gross structure of Highland Britain in the west and Lowland Britain in the east is entirely a product of the North Atlantic plate tectonic events. The last great tectonic event, the Alpine uplift of the Tertiary, imposed the latest trends. Strong E–W drape folding, inversion of sedimentary basins, emphasis of the central Irish Sea and Cheshire Rifts, and rejuvenation of the basement structures and uplift, imposed the

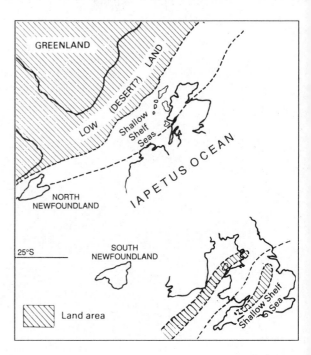

Fig. 3. The Iapetus Ocean showing the basement rocks of 500 million years ago.

(*Source*: Brunsden *et al.* 1988, fig. 2.10.)

present relief of Britain as well as the drainage directions.

Throughout this whole period marginal basins controlled sedimentation. The Rockall Trough, Porcupine Seabights, Irish Sea, and North Sea were major features. On a smaller scale successive centres of deposition occurred but as time passed and the structural pattern developed several important lines became dominant. The London Basin, Hampshire Basin, Celtic Sea, Bristol Channel, Cardigan Bay, and Lough Neagh, are typical and recurrent examples. On a still smaller scale the Tremadoc Bay, Lundy, Petrockstow, and Bovey Basins became important centres. Today these form the visible 'outline' of Britain and remain centres of sedimentation, tectonic movement and unrest.

Another implication of the plate tectonic location of Britain is that we have not remained in the same latitudinal–longitudinal position during the course of our

evolution. Not only has our location changed from being in the middle of a supercontinent to a marginal ocean-basin position, but also we have moved north. It is often difficult to understand how a temperate country could once have been tropical or desert. Plate tectonics makes this easier to accept. In 500 MA we lay in the southern hemisphere at latitude 45° and separated into two halves on either side of an ocean. We drifted north and west, joined together on a suture line and were again uplifted. By 370 MA we were eroding in a southern desert. Crossing the equator and covered in tropical forests (310 MA) we then passed north to the deserts of the Permo-Trias and toward our present position, travelling at around 11 km a million years. Finally, things began to get cold as we moved into northern waters and the continental position allowed the growth of ice sheets.

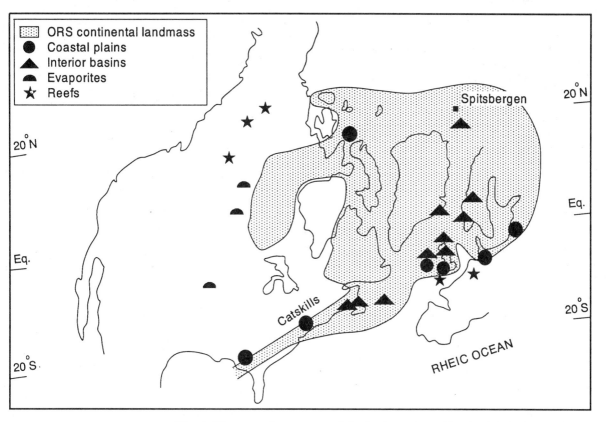

Fig. 4. The approximate extent of the landmass created by the Caledonian uplift as it may have appeared in Old Red Sandstone (Devonian) times. Note the latitude. The British Isles were south of the Equator and the climate was arid.

(*Source*: Anderton *et al.* 1979, fig. 9.1.)

The Caledonian land mass

Approximately 500 MA the small islands that were to become the British Isles lay within an early ocean, perhaps 2000–4000 km wide, known as the Iapetus Ocean. The basement rocks of northern Scotland lay within a shallow shelf sea attached to Greenland and Canada. England and Wales lay on the other side of the Ocean. About 400 MA movements of the continental plates began to close the ocean at a rate of perhaps 0.4–0.5 cm a year until the two sides of the ocean welded together along what is called a suture line from Solway to Shannon.

It is believed, contrary to popular opinion, that the compression was vice-like and fairly gentle and episodic. It did *not* produce towering mountain ranges and structures like those of the Alpine–Himalayan ranges today. Instead it produced at first shelf-and-basin continental margins, volcanism and, as the ocean closed, a broad continental landmass covering the then closer together, Scandinavia, Greenland, northern Canada, Appalachians, and Britain.

There is evidence that the concluding phase showed an increasing pase of uplift, and certainly high mountains existed because coarse-grained deposits were soon forming in the subsiding basins of the Devonian period (Figure 5). Here were formed the red desert rocks and coral reefs of Wales and Devon, the Midlands, the Border Basin of Northumbria, the Midland Basin of Scotland, and Orcadia.

These episodes effectively began the construction of the landscape we know today producing the fabric of the Southern Uplands of Scotland, the Lake District, Isle of Man, northern Wales, and eastern Ireland. Beneath the surface the granites of Dumfries and Wicklow were emplaced. Some rocks were changed by heat and pressure, metamorphosed to form rocks like the Blaenau Festiniog slates, and lavas were extruded as the Borrowdale Volcanics. It was a typical subduction margin of continental collision.

The whole area had a NE–SW alignment due to the orientation of the mountains, and great faults, like the strike-slip Great Glen Fault, were formed and still dominate the structure of today's landscape.

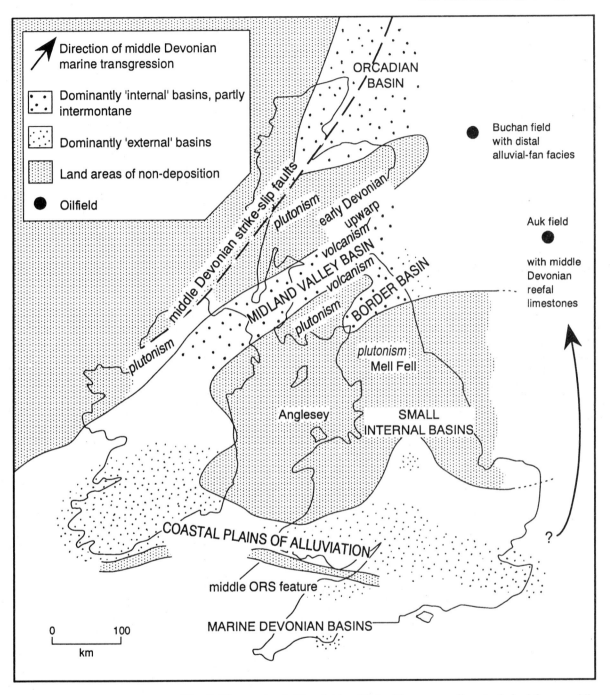

Fig. 5. The great subsiding basins filled with coarse sediments derived from rapidly uplifting mountains about 400 million years ago.

(*Source*: Anderton *et al.* 1979, fig. 9.14.)

Hercynides

The development of depositional basins led to an exciting sequence of geological events (Figure 6). The southern margins of the Caledonian continent began to subside and a major marine invasion took place. From what is now eastern USA to Northumberland and the Rhine, basins and blocks jostled up and down to allow the amazingly varied Carboniferous deposits of carbonate seas, deltas, and organic swamps which became the upland limestone massifs, moorlands, and coalfields of today (Figure 7). The area eventually became an enormous fluvio-deltaic plain covering much of north-west Europe.

The next dramatic episode was about to begin. Far to the south the restless movement of the African continental plate seems to have created a subduction zone, in what is called the Rheic Ocean, with marginal basins across the Rhine and massive uplift, volcanism, and granite emplacement. There were great strike-slip faults across Spain and France and in the structural fabric of southern Britain (Figure 6).

It is believed that finally a block-faulted mountain range stretched from Central Europe through Dartmoor to eastern USA, which was, of course, much closer at that time. The granites, tin, and copper of south-west England, and the minerals of the Pennines and central Ireland were emplaced. To the north Britain was now divided into major units like the Pennines and many Caledonian structures were strengthened. This northern area was a much denuded remnant of a mountain mass.

Between the Caledonian and Hercynian regions large structural blocks separated great sedimentary basins. The Bristol Channel, Severn, Cheshire, Irish Sea, North Sea, and Cardigan Bay basins were taking shape. Beneath what was to become southern Britain structural blocks were beginning to be buried in softer strata. The climate was again desert, and the surface a hostile one of dunes, alluvial fans, braided streams, and saline lakes. The Permo-Triassic had arrived and the stage was set for the development of the landscape we know today. Britain lay in the centre of the newly formed Pangaea supercontinent.

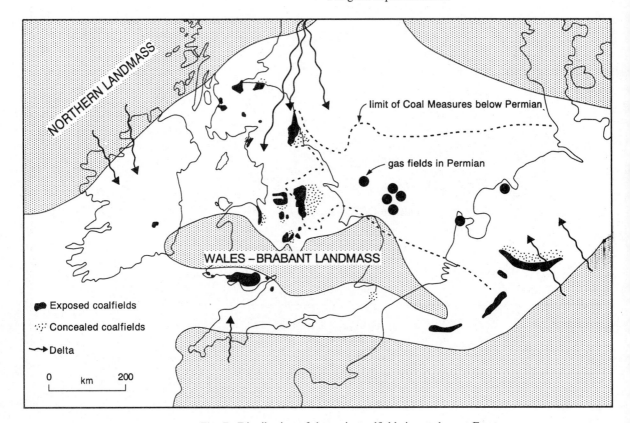

Fig. 7. Distribution of the main coalfields in north-west Europe.
(*Source*: Anderton *et al.* 1979, fig. 11.18.)

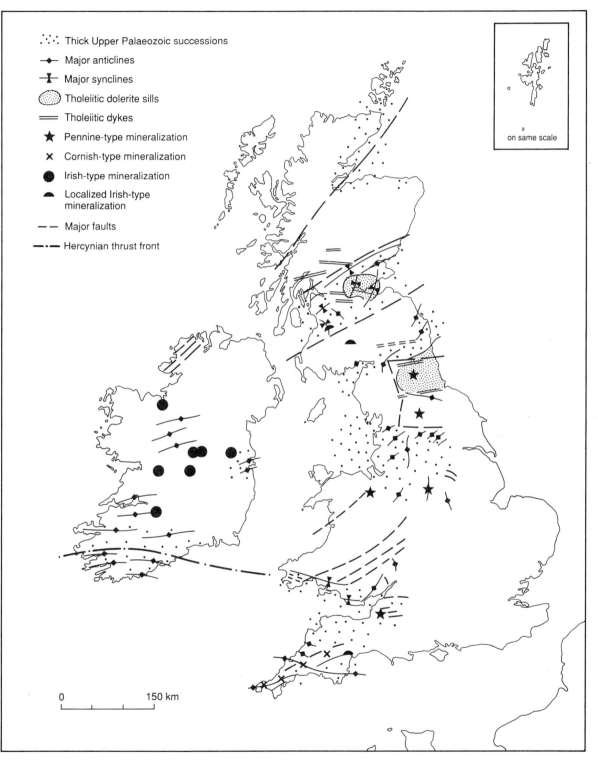

Fig. 6. Important structural features of the Hercynian orogeny.
(*Source*: Anderton *et al.* 1979, fig. 12.12.)

Mesozoic and Cainozoic events

Following the extreme desert geomorphological conditions of the Triassic the British region underwent a complex series of uplifts and transgressions of the sea that during the Jurassic and Cretaceous periods largely created the Mesozoic rocks of lowland England. All of this is related to the startling fact that prior to 200 MA the North Atlantic Ocean did not exist. Instead the continental plates of western Europe, Rockall, Greenland, and North America formed the single plate of Pangaea.

The division began in the middle Jurassic with rifting in the North Sea basin and major sea-floor spreading between Ireland and Greenland. A very complex series of basins, horsts, grabens, and rifts mark this dynamic margin and the main events are summarized in Figure 8a–e.

During the late Mesozoic Era something again began to stir in the south (Figure 8d). Great compressional forces heralded another orogenic phase—a collision between Africa and Europe. The movements climaxed in the late Eocene and Oligocene. The Alps are the main product of this activity, with intensely deformed thrust sheets and nappes, gravity sliding, and intense marginal erosion (Figures 8–10).

The whole region was in tension as the continent pulled apart (Figure 9) and the ensuing crustal extension released floods of volcanic activity for nearly ten million years. Flood basalts up to 2000 m thick were created in Mull and Antrim, as were the Lundy granites, and the massive dyke swarms which extended from Scotland to Snowdonia (Figure 11).

This uplift was accompanied by vigorous sea-floor spreading in the Atlantic, and the two effects were magnified in Britain to give the contrasting effects of block movement in the ancient Caledonian, Hercynian, and

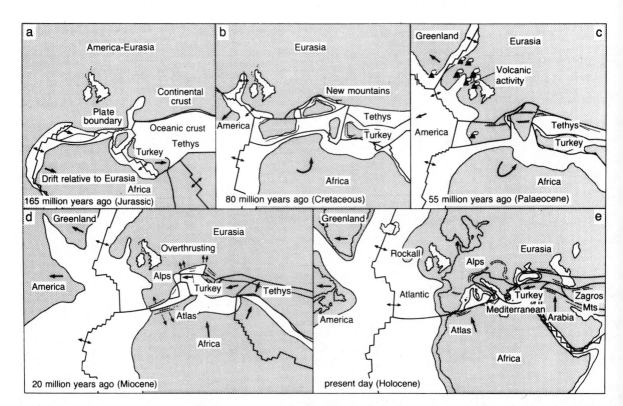

Fig. 8. The plate tectonic evolution of the Mediterranean basin and the North Atlantic region since the Jurassic. Note the relatively late opening of the North Atlantic and the associated volcanic activity in the Palaeocene.

(Modified after Dunning *et al.* 1978, fig. 93.)

basement structures, sediment draping over the block units, true E–W folding and the development of N–S, NW–SE fault systems and graben basins like the Sticklepath–Lustleigh Fault. It is now believed that many of the land surfaces, soils, and drainage patterns of Britain have to be explained in terms of the unstable tectonic history rather than the traditional view of tectonic stability.

The Miocene period is marked by uplift and the Tertiary basins show increasingly coarse sediments. Middle Oligocene beds are folded and then cut by Pliocene surfaces and there is a major lack of Miocene sediments onshore because uplift was taking place.

Throughout the Neogene (after 26 MA) and Pleistocene active subsidence was taking place in the North Sea basin. There is over a kilometre of Neogene

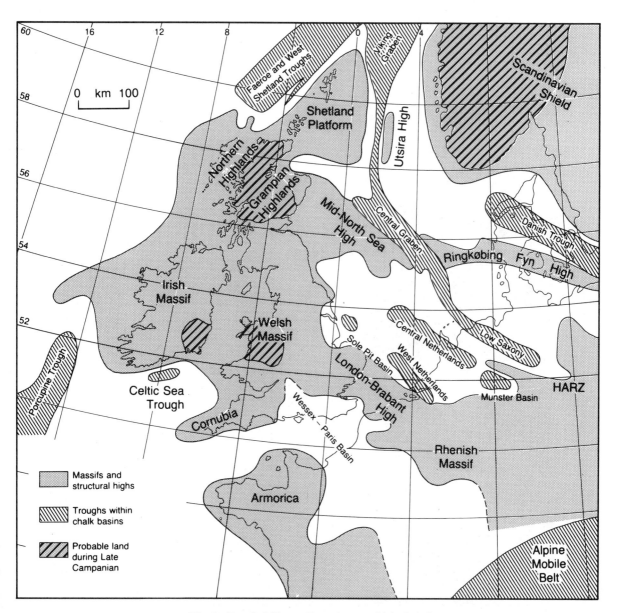

Fig. 9. Detailed Upper Cretaceous geological setting.
(*Source*: Goudie, 1990, fig. 1.26.)

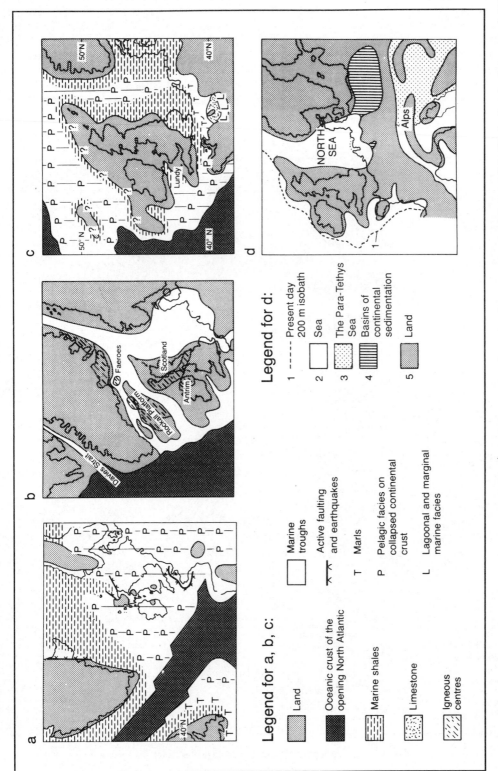

Fig. 10. The palaeogeography of the British Isles and environs at various stages from the Cretaceous to the Miocene. (Modified after Anderton *et al.* 1979 figs 15.15, 16.4, 16.8 and 16.10.) (*a*) Generalized palaeogeography in the Upper Cretaceous (Campanian); (*b*) Palaeocene palaeogeography. Note the position of the British Isles with respect to Greenland and the widespread nature of igneous activity; (*c*) Eocene palaeogeography, key as for (*a*); (*d*) Miocene palaeogeography. 1 = present day 200 m isobath; 2 = sea; 3 = the Para–Tethys sea; 4 = basins of continental sedimentation; 5 = land.

sediment in the centre (Figure 12) with large quantities coming from the uplifting Alps via the Rhine. These fluctuating conditions continued right up to the Late Pleistocene (Figure 10). Indeed there is evidence that subsidence and tectonic movements continue today, though they are complicated by the isostatic effects of the glacial unloading and the subsequent recovery of Scotland.

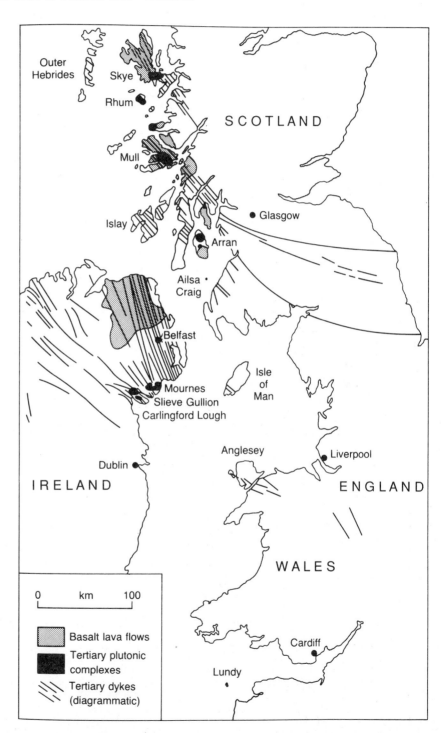

Fig. 11. The opening of the North Atlantic produced a major spasm of igneous activity in the British Isles leading to the intrusion of plutons (including the Isle of Lundy), the injection of swarms of dykes, and the extrusion of expansive basalt lava flows.

(*Source*: Goudie, 1990, fig. 1.10.)

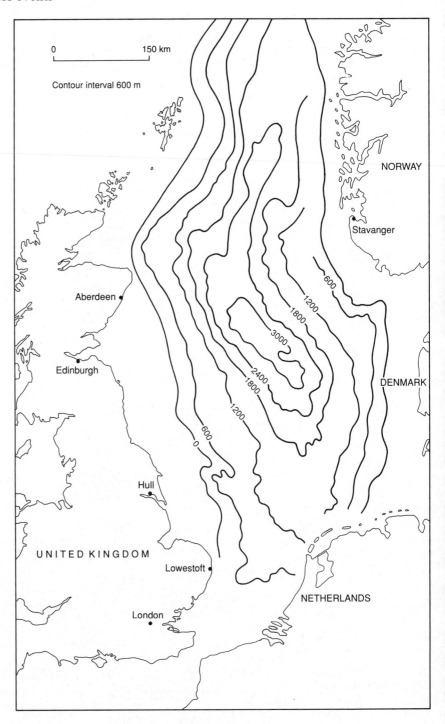

Fig. 12. The subsidence of the North Sea basin during the Tertiary and Quaternary has enabled the accumulation of large thicknesses of sediment on top of the chalk. In the case of the central part of the basin there are over 3000 m of material.

(Modified from Lovell 1986, fig. 8.3.)

Rocks

The rich and varied events which have shaped the British Isles, the marginal position between tectonic provinces, the changing latitudinal position, fluctuating sea-levels, and changing climates have brought together an almost unrivalled variety of rocks in a very small space. Episodes of geological history representing the last 1000 million years can be studied and the oldest rock known in Britain is 2700 million years old. It is therefore only possible in a short summary to record the basic patterns.

The British Isles can be divided into two parts along the Tees–Exe line (Figure 13). To the north and west

Fig. 13. Mackinder's division of Britain into areas underlain by ancient and new rocks.

0 150 m

■ Surface of Carboniferous and older rocks

□ Surface of rocks newer than Carboniferous, underlain by Carboniferous and older rocks

are the old, more resistant rocks of Highland Britain. These are Pre-Carboniferous in age and aligned with the Caledonian and Hercynian mountain trends. They are block-faulted and divided by major strike-slip faults. Typical rocks are the 'ancient lands' of the Pre-Cambrian basement in the Scottish Highlands, Western Isles, Ulster, Donegal, and Mayo; the gneiss of Lewis and Sutherland; the granite of the Cairngorm; and the quartzite and Torridonian sandstones of Sutherland (Figure 14). Occasionally, too the basement pokes up to form the Malvern Hills, Longmynd, Charnwood Forest, Anglesey, Lleyn peninsula, Rosslare, Nuneaton,

Ingleborough, and the Lizard, but these are rare glimpses of a very complicated and ancient story.

A little to the south lie the Caledonian terrains of the Southern Uplands, Lake District, Isle of Man, Wales, and Ireland. They are composed of sandstones, limestones, and clays. Often they have been metamorphosed: changed by heat and pressure to form quartzites, marbles, and slates. Set within them are the granites of Wicklow and Dumfries. Economically useful rocks include the Blaenau Festiniog, Wenlock, and Tremadoc slates, many of which include early fossils like the trilobites and graptolites (Figure 15).

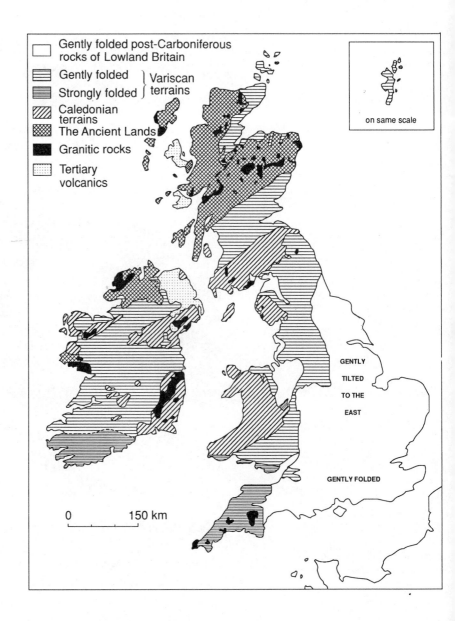

Fig. 14. The fundamental fourfold division of the structure.

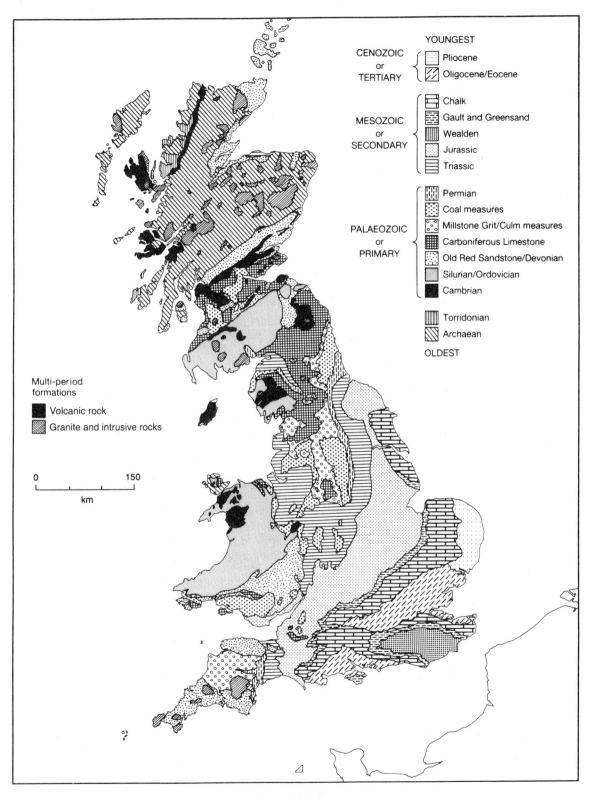

Fig. 15. Solid geology of Britain.

(Simplified from Jones and Mattingley 1990.)

The legend contains:

YOUNGEST

CENOZOIC or TERTIARY
- Pliocene
- Oligocene/Eocene

MESOZOIC or SECONDARY
- Chalk
- Gault and Greensand
- Wealden
- Jurassic
- Triassic

PALAEOZOIC or PRIMARY
- Permian
- Coal measures
- Millstone Grit/Culm measures
- Carboniferous Limestone
- Old Red Sandstone/Devonian
- Silurian/Ordovician
- Cambrian

- Torridonian
- Archaean

OLDEST

Multi-period formations
- Volcanic rock
- Granite and intrusive rocks

0 — 150 km

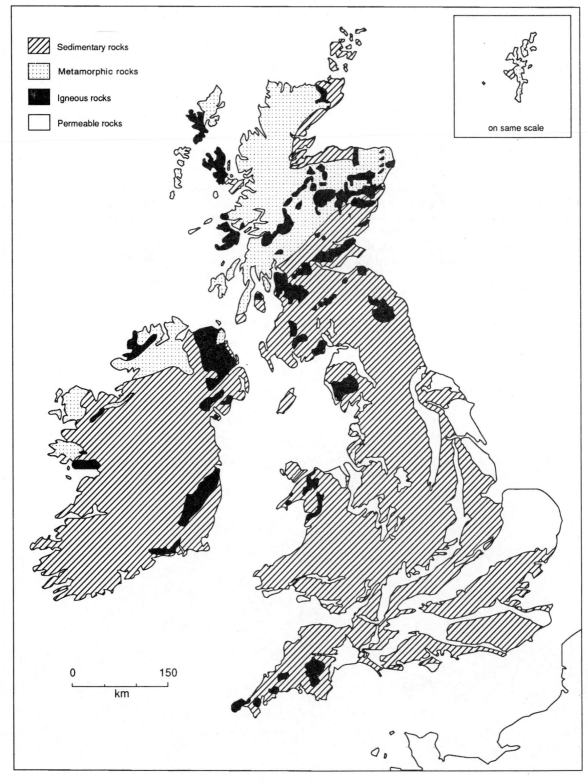

Fig. 16. The main areas of sedimentary, metamorphic, igneous, and permeable rocks.

Surrounding these ancient mountains are the products of their decay: thick desert gravels in central Scotland and Herefordshire; coral reefs and ocean muds in Devon; then the limestones, delta grits (Millstone grit), and Coal Measures of the Carboniferous swamps. Finally there are the metamorphic rocks, minerals, and south-western granites of the Hercynian uplift. Again these have lovely names, like the Baggy Beds, and the Hangman Grits, and a vast proliferation of plant and animal life.

Once more Britain lay within the hinterland of a supercontinent, Pangaea. The rocks that were formed were the desert products of the denuding Hercynian mountain range and so, as we step across the Highland–Lowland boundary, we move to the younger, relatively weak sedimentary rocks derived from these ancient land masses (Figure 16).

At first they are the clastic, gravel, sandstone, and saline rocks of the Permo-Triassic. These form the red rocks of the South-West, Cheshire, and the Midlands.

Overlying them are the wonderfully fossiliferous Liassic and Jurassic silts, clays, and limestones that form some of the most beautiful building stones of Britain (e.g. the Hamstone), the Cotswold escarpment, and the great clay vales of Kimmeridge and Oxford. The beauty of the spectacular and varied coast of Dorset depends on these rocks. Further east the clays, sands, and chalk of the Cretaceous and then the mixed sand, gravel, and clay deposits of the Tertiary form the landscapes of the Weald, Hampshire Basin, and the scarplands of the Salisbury Plain, Chilterns, North and South Downs, and the Lincoln Edge.

Finally and spread widely as far south as London are the deposits of the last great landscape influence—ice. Britain also lay at the climatic change crossroads. Across the ancient tropical landscape of the Tertiary Period marched the great Pleistocene ice sheets. Tills, gravels, outwash, and wind-blown deposits complete the scene.

Relief, relief types, and relief regions

Although a simple primary division can be made between Highland and Lowland Britain the main characteristic of Britain is that it is so varied in relief, colour, texture, and character (Figure 17). This is, of course a reflection of the underlying rocks and structures but it is also a product of the varied climates, environments, and processes of the last 50 million years.

The reasons for the variety in relief include the structural history which has given a very irregular boundary to the highland limit; the incidence of subsiding basins divided by mobile faulted blocks; the great variety of rock types of different resistance to erosion; the intrusions of granitic rocks which form large relief masses; the variable uplift (or subsidence) histories of the different regions, and the complexities of the processes which have denuded them at different rates. The glacial erosion of deep valleys in the north, and the coincidence of this with the highest available elevations and toughest rocks, gives maximum relief. Conversely, the deep incision of rivers to low Pleistocene sea-levels and the less resistant (to fluvial erosion) rocks in the south, give deep valleys in quite a low-lying area of level erosion surfaces. Each area has to be understood in terms of its own history. Broadly, however, it can be shown that elevations and relative relief values decline from mountains and plateaux over 600 m in Scotland and Wales, to high plateaux and dissected hills at 200–600 m in the Southern Uplands, northern and southern Ireland, the Pennines, North York Moors, central Wales, and Dartmoor. East and south of this the land drops quickly away to the low plateaux of the Midlands, the scarplands, at about 100–200 m, and the low country of the Fens. Central Ireland too is a low depression. The pattern is not simple because low land follows the coast, the structural depressions, major fault lines, and clay vales. The influence of the structural history is very clear in these areas.

The variety of rocks and the denudational history makes it possible to divide the country into quite small distinctive regions. They often have sharp, geologically defined boundaries and sudden changes from one landscape to another. Each is a palimpsest of processes and geological events covering millions of years (Figure 18).

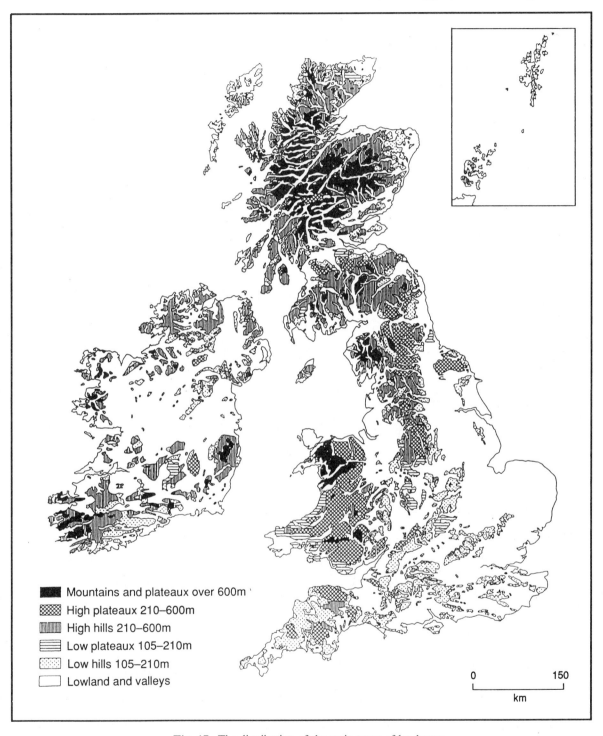

Fig. 17. The distribution of the main types of landscape.

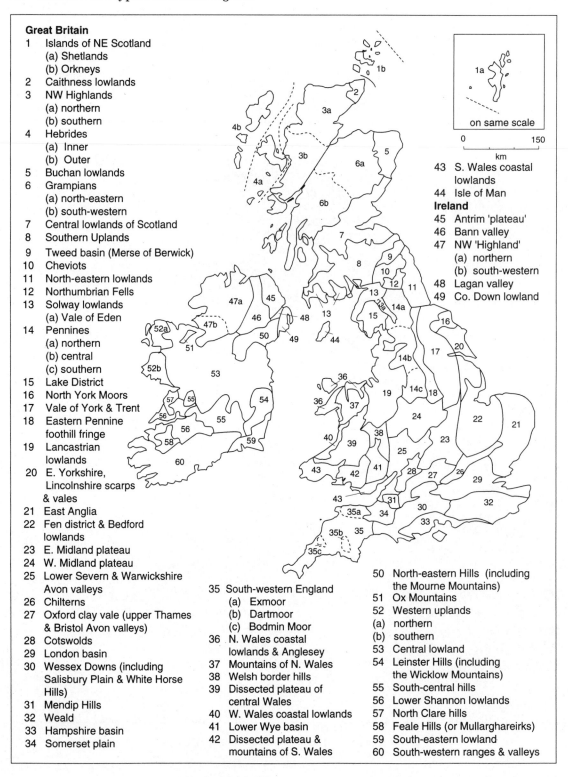

Great Britain
1   Islands of NE Scotland
    (a) Shetlands
    (b) Orkneys
2   Caithness lowlands
3   NW Highlands
    (a) northern
    (b) southern
4   Hebrides
    (a) Inner
    (b) Outer
5   Buchan lowlands
6   Grampians
    (a) north-eastern
    (b) south-western
7   Central lowlands of Scotland
8   Southern Uplands
9   Tweed basin (Merse of Berwick)
10 Cheviots
11 North-eastern lowlands
12 Northumbrian Fells
13 Solway lowlands
    (a) Vale of Eden
14 Pennines
    (a) northern
    (b) central
    (c) southern
15 Lake District
16 North York Moors
17 Vale of York & Trent
18 Eastern Pennine
    foothill fringe
19 Lancastrian
    lowlands
20 E. Yorkshire,
    Lincolnshire scarps
    & vales
21 East Anglia
22 Fen district & Bedford
    lowlands
23 E. Midland plateau
24 W. Midland plateau
25 Lower Severn & Warwickshire
    Avon valleys
26 Chilterns
27 Oxford clay vale (upper Thames
    & Bristol Avon valleys)
28 Cotswolds
29 London basin
30 Wessex Downs (including
    Salisbury Plain & White Horse
    Hills)
31 Mendip Hills
32 Weald
33 Hampshire basin
34 Somerset plain

35 South-western England
    (a) Exmoor
    (b) Dartmoor
    (c) Bodmin Moor
36 N. Wales coastal
    lowlands & Anglesey
37 Mountains of N. Wales
38 Welsh border hills
39 Dissected plateau of
    central Wales
40 W. Wales coastal lowlands
41 Lower Wye basin
42 Dissected plateau &
    mountains of S. Wales

43 S. Wales coastal
    lowlands
44 Isle of Man
Ireland
45 Antrim 'plateau'
46 Bann valley
47 NW 'Highland'
    (a) northern
    (b) south-western
48 Lagan valley
49 Co. Down lowland

50 North-eastern Hills (including
    the Mourne Mountains)
51 Ox Mountains
52 Western uplands
    (a) northern
    (b) southern
53 Central lowland
54 Leinster Hills (including
    the Wicklow Mountains)
55 South-central hills
56 Lower Shannon lowlands
57 North Clare hills
58 Feale Hills (or Mullarghareirks)
59 South-eastern lowland
60 South-western ranges & valleys

Fig. 18. Major relief regions.

(*Source*: Warwick, fig. 13, in Watson and Sissons, 1964.)

The origin of drainage

One of the fundamental Mesozoic geological events was the deposition of the chalk during the Upper Cretaceous transgression. It is believed that this virtually covered Britain in Albian times and that this pure limestone accumulated in the basins around the highlands to a thickness of up to 500 m.

It is probable that the transgression caused marine erosion in the underlying rocks as it stepped unconformably across Britain. For this reason it is often said that it sealed off the past. If we wish to understand the present landscape we must begin therefore with the emergence of a new landscape from beneath the Cretaceous Sea.

As the sea retreated as a result of local tectonic uplift and eustatic (world-wide) sea-level changes, rivers began to flow across the new surface. The subsiding and developing basin of the North Sea was one obvious control as were the other Tertiary sedimentary basins. Many

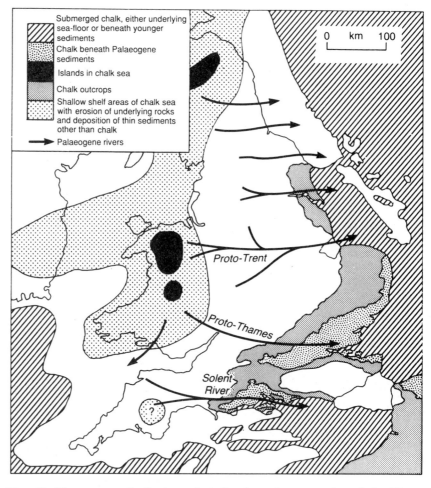

Fig. 19. The present distribution of chalk, the palaeogeography of the Upper Cretaceous (chalk) sea, and the presumed courses of the Palaeogene consequent drainage pattern that developed on the newly emerged chalk. Areas where chalk is thought to have been deposited but subsequently removed by denudation are shown in white.

(Modified after Jones 1985, fig. 1.15.)

geologists believe therefore that the early drainage pattern was from the highlands of the west toward the Sole Pit Basin between the Mid-North Sea High and the London–Brabant High. The Wessex– Paris Basin and the Celtic Sea trough were other 'sinks' for water and sediment (Figure 19).

One consequence of this hypothesis is the idea that the chalk cover has been stripped from large areas of Britain and that the early drainage pattern has been superimposed on to the underlying rocks. Although there are few signs of weathered remnants of the chalk cover, support for the idea comes from the fact that many of today's rivers are discordant to the underlying structures. This is particularly true of the Welsh rivers which cross many of the Hercynian and Caledonian structural lines, such as the south Wales synclinorium.

If this hypothesis is true it is also necessary to accept that the early consequent streams such as the proto-Trent, proto-Thames, and the Solent River have since suffered major adjustment, river capture, diversion, and disruption. This would be necessary if, for example, the Severn, Trent, or Thames catchments were to develop to their present form.

It also has to be accepted that the present-day landscape of Britain is largely a creation of the Tertiary period in which the main outlines of the mountains, basins, valleys, and coastline were developed. How this happened is one of the great detective stories of British Science.

Seismic activity and tectonic deformation

The seismic and tectonic activity of the Miocene and Pleistocene continues today. Although the present-day seismicity is low, minor fault movements do occur, such as the Sticklepath Fault movement of 1954, and earthquakes are common. The shocks are, however, small and infrequent. For example, 2000 events have been recorded since AD 1185. No known earthquake has exceeded a magnitude of 5.5 and maximum epicentral intensities have not exceeded VIII.

The map of known epicentres (Figure 20) emphasizes certain long-established geological trends. The Great Glen Fault of the Caledonian orogen still records tremors of 4.5 magnitude. The south Wales–Herefordshire line has many tremors. One third of all British earthquakes with a magnitude of greater than 4 are located near the Neath–Swansea disturbance and the damaging earthquake of 1896 at Hereford follows this trend. Events related to the North Sea tectonic structures and subsidence are well marked, with events in 1927 and 1931 east of the Humber, 1382 in north Kent, and 1580 in the Straits of Dover (the so-called London 'quake) ranking among the biggest in the British Isles. The largest so far recorded is the Colchester event of 1884 which was felt over 200 km away, killed 3 people, and damaged 1000 buildings. Its magnitude was only 4.4 but it was shallow and therefore damaging. Most of the small tremors in Britain are associated with coal mining, subsidence, and reservoir filling. There tends, therefore, to be complacency but it is worth pointing out that prediction of big events is not possible when only events up to magnitude 5.5 have been recorded.

A useful complementary map is one of estimated current rates of crustal movement (Figure 21). It is difficult to get accurate figures and to separate these from sea-level change. The available survey data, and estimates from the uplift of datable deposits such as raised beaches, do allow a general picture to emerge.

Scotland is rising at rates between 0.5 and 2.0 mm/y due to isostatic recovery from the release of the weight of the ice sheets of the last glaciation and the denudation of great thicknesses of rock from the glaciated valleys. South-east England appears to be sinking at rates up to 2.0 mm/y with the neutral line running from Middlesborough to north Wales. The South-east pattern is that of a basin subsidence, perhaps related to the sediment loading of the River Thames discharge but mainly due to the very long continued North Sea subsidence. South-west England and Wales also appear to be subsiding—at least on the coast. The south coast is relatively stable.

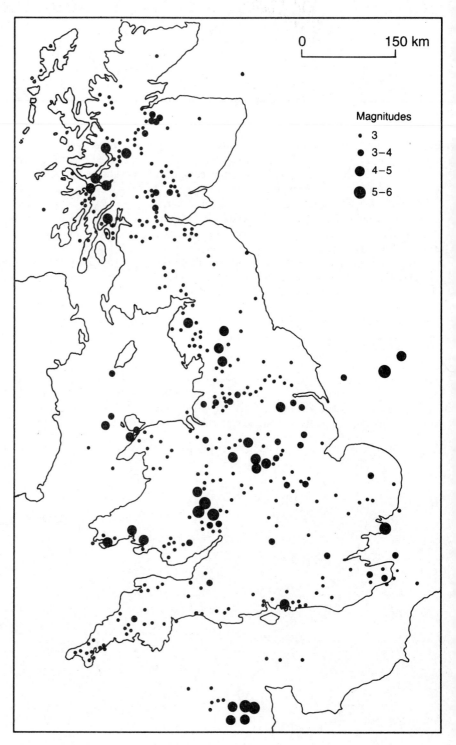

Fig. 20. Known earthquakes based on analysis by British Geological Survey.

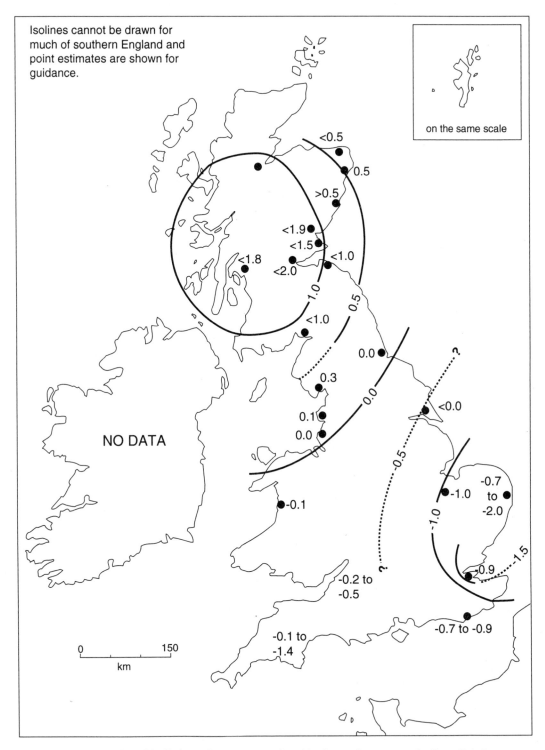

Fig. 21. Estimated current rates (mm/y) of crustal movement in Great Britain.
(From Shennan 1989, fig. 9.)

Geomorphology

The glaciation of Britain

Approximately 2.4 MA, in the late Pliocene, full glacial climatic conditions afflicted the British Isles, which prior to that time had enjoyed the relative warmth of the mid- and early Tertiary. The following 2.4 million years were a time of remarkably extreme, frequent, and rapid climatic changes, during which a whole suite of cold episodes (glacials) and warmer episodes (interglacials) occurred. The warm interglacials had climates which were broadly comparable to that of the present day. They were relatively short-lived, only constituting about 10% of the Quaternary era (which started about 1.6 MA, and which witnessed around 17 glacial/interglacial cycles). Similarly, the full glacial conditions only existed for a short time. Conditions were rarely stable. The key word for the Quaternary is *change*.

The evidence for such multiple changes of climate is better preserved in the sediments of the ocean floor than it is on land, where later glaciations have largely removed the evidence of former events. Indeed, on land there is only clear evidence for a few glacial advances, and their sequence, dating, and extent are still the subject of controversy.

There is evidence on land for the Anglian glaciation, which probably occurred about 450 000 years ago. Its extent is delimited by dispersed scatters of glacial debris (till) which has suffered much post-depositional erosion and weathering. The Anglian advance may have been the most extensive to affect the British Isles, reaching as far south as Oxford and south Essex. At the maximum extent of glaciation ice reached as far as the Scilly Isles, the Isle of Lundy, and the north coast of the south-west peninsula (Figure 22).

The last major glacial event probably started round about 110 000 years ago, and is called the Devensian. Devensian ice reached its maximum extent at about 17 000 years ago (during the so-called Dimlington stadial) but was much less extensive than in the Anglian. None the less it was in places over 1200 m thick, and perhaps over 1800 m thick over highland Scotland. It covered much of Wales, Scotland, Ireland, and northern England. Parts of southern Wales and the far north-east of Scotland may not have been covered. The ice touched the Yorkshire coast and the extreme north of Norfolk, but did not succeed in advancing very far into the Midlands. It is also probable that it did not link up with Scandinavian ice.

The ice had largely disappeared by 13 000 years ago (Figure 23), as indicated by the dates for sediment that accumulated in lakes following ice-sheet wastage, but cold conditions led to a short-lived glacial re-advance—the Loch Lomond stadial—at about 11 000 years ago. Glaciers re-formed in many highland areas, notably in the Scottish Highlands, but also in the Lake District, north Wales, and the Brecon Beacons. The end moraines produced by this event are still clearly displayed in the present landscape. During all the glacial advances southern England lay beyond the limit of glaciation, and so was subjected to the full rigours of a tundra environment.

Table 1. Zones of glacial erosion

Zone	Lowlands	Uplands
0	No erosion. Head on weathered rocks and slopes. Outwash in concavities. Rare occurrences of till on weathered rock.	No erosion. Outwash on valley floors. Solifluction deposits on slopes, boulder fields and tors on divides.
I	Ice erosion confined to detailed or subordinate modifications. Concavities drift-mantled but convexities may show some ice moulding. Occasional *roches moutonnées*. Ice-scoured bluffs in favourable locations.	Ice erosion confined to detailed or subordinate modifications. Suitable valley slopes ice-steepened. Entrenched meanders and spurs converted to rock knobs. Interfluves still commonly Zone 0.
II	Extensive excavation along main flow-lines so that concavities may be drift-free or floored by outwash or post-glacial deposits. Isolated obstacles may be given ovoid or cutwater forms if of soft rock, or crag-and-tail with associated scour troughs if the rock is hard.	Conversions of preglacial valleys to troughs common, but usually confined to those of direction concordant with ice flow. Some diffluence; transfluence rare. Interfluves may be Zone I or even Zone 0, and separated from troughs by well-marked shoulders.
III	Preglacial forms no longer recognizable but replaced by tapered or bridge interfluves with planar slopes on soft rocks, and by rock drumlins and knock-and-lochan topography on hard rocks.	Transformation of valleys to troughs comprehensive giving compartmental relief with isolated plateau or mountain blocks. Zone 0 may still persist on interfluves at sufficiently great heights.

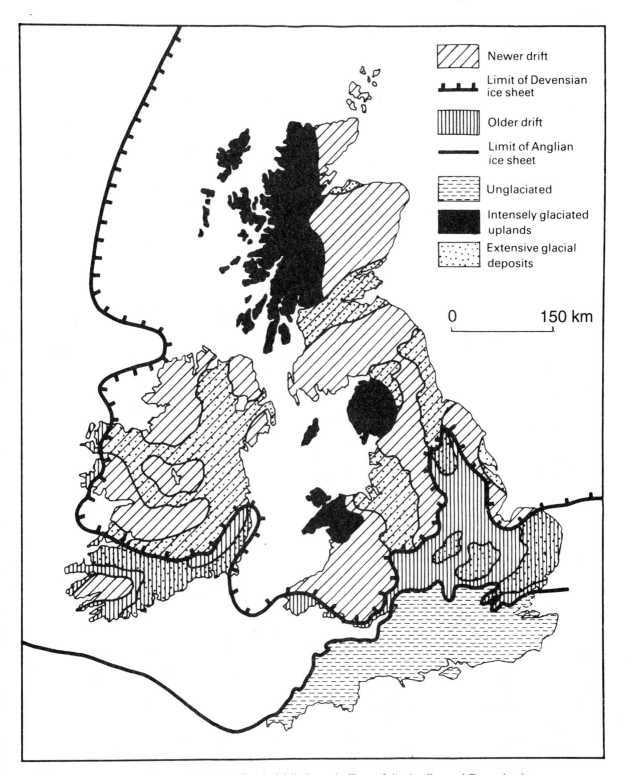

Fig. 22. The glacial limits and effects of the Anglian and Devensian ice.

The distribution of glacial erosion

In the various glacial advances that occurred in the British Isles during the Pleistocene, the glaciers and ice sheets achieved differing degrees of landscape modification in different parts of the country. It is possible to recognize five empirically derived categories from Zone 0, which is virtually unmodified, to Zone IV, where the entire landscape has been shaped by ice. The characteristics of these different zones are shown in Table 1.

As one might expect, given their high relief and plentiful precipitation, the zones of highest erosional activity include the western parts of the Scottish Highlands, the Lake District, and north Wales. These are all areas where there are major cirques, deep glacial troughs, scoured surfaces with striations and *roches moutonnées*, examples of glacial diffluences, and severe drainage modification (Figure 24).

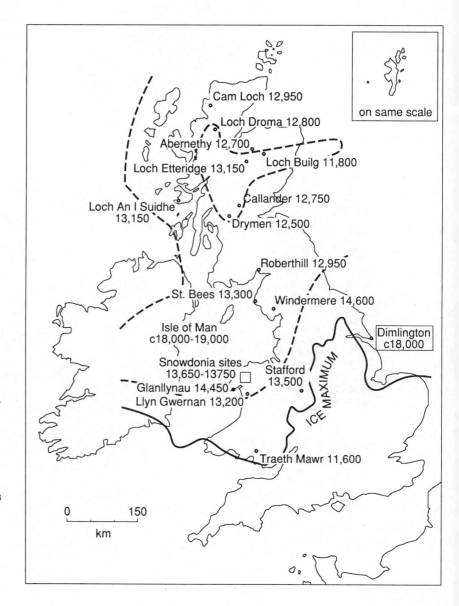

Fig. 23. The extent and decay of the Devensian Glacial, showing the probable maximum ice thickness at the time of the maximum glaciation expressed as ice surface contours. Also shown are radio-carbon in years BP dates from the earliest organic sediments that accumulated in lakes following the wastage of the Late Devensian ice sheet.

(*Source*: Modified after Lowe and Walker 1984, fig. 7.13.)

However, it would be wrong to dismiss the role of glacial erosion in low-lying areas, even though glacial deposition seems at first sight to be more important in areas such as East Anglia. The composition of the till that makes up many of the deposits indicates that much of it was derived from the erosion of susceptible materials, including the chalk, and the clays of the Lias, Jurassic, and Cretaceous. Glacial erosion planed off the chalk escarpment in East Anglia, causing it to be very subdued in comparison with that of the Chilterns further south. Glacial erosion by ice moving into the

Cheshire–Shropshire plain dug deeply into the area now occupied by the Dee and Mersey estuaries.

The ice-flow lines for the Devensian and Anglian glaciations show a close correspondence with the zones of maximum erosion. The pattern reflects several controls. There is an obvious relationship to the flow along the steep ice-thickness gradients. Strong erosion also occurred on the troughs between the ice dispersal centres as well as below the main advancing lobes. Locally rock type determined the differential rates.

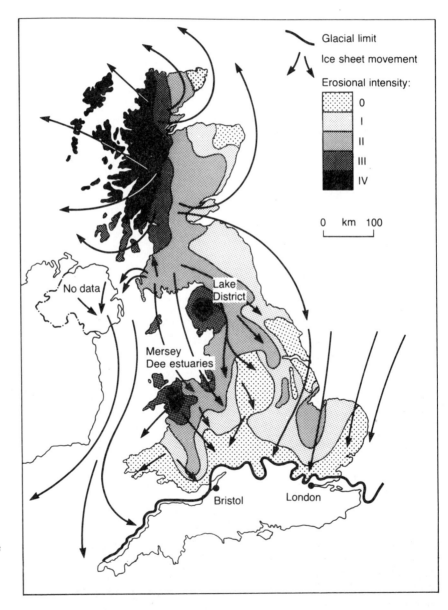

Fig. 24. Patterns of ice-sheet movement and glacial erosion. For explanation of the five zones of erosional intensity see Table 1.

(Modified from Jones, 1985, fig. 1.23.)

Glacial deposition

The glaciers deposited large quantities of material, which includes till, supraglacial deposits, and outwash (Figure 25). Conventionally a distinction is drawn between those materials laid down in the Devensian Glaciation and those laid down in earlier glacial phases. The former, called the 'Newer Drift', have the clearest morphologies, and it is possible to identify relatively fresh depositional landforms (e.g. kames, eskers, kettle holes, drumlins, or moraines). By contrast, the latter types of deposit, called the 'Older Drift', which lies to the south of the Devensian limit, have been so moulded by subsequent events, especially periglacial activity, that the former presence of ice is primarily indicated by rather formless, residual spreads of till and outwash gravel. Indeed, some of the earliest glacial materials, including the Plateau Drift of the Oxford region, are often little more than scattered patches of pebbles.

The nature of the deposits varies according to their area of derivation. Thus Scandinavian erratics are found in eastern portions of Britain, while in western and mid-

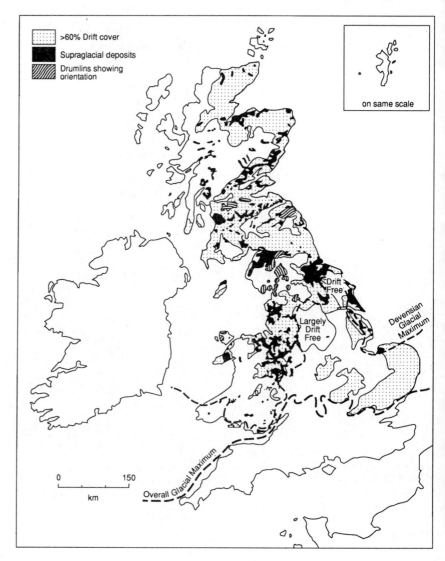

Fig. 25. The extent of glacial deposits, drumlins, and supraglacial sediments within the late Devensian ice-sheet limit.

(After Boulton *et al.* 1977, fig. 17.4.)

land England, there are large quantities of material that were derived from ice that came across the Irish Sea. Along the shores of the Irish Sea there are numerous erratics that are derived from the highly characteristic outcrop of riebeckite-eurite that comes from the Scottish island of Ailsa Craig.

One particular type of glacial deposition feature that is very widespread is the drumlin, an elongated hill that formed beneath the ice (Figure 26). All such drumlins occur within the limits of the Devensian 'Newer Drift', and especially large concentrations occur in the north and west of Ireland, in parts of north Wales, in the lowlands surrounding the Lake District and the Pennines, and in the Central Lowlands of Scotland. This distribution is related to areas of deposition of thick subglacial till which have not subsequently been covered by supraglacial deposits or proglacial outwash.

The term 'supraglacial' refers to materials laid down from the surface of melting and decaying glacier ice. The largest area of such deposition in Britain occurs on the west flank of the Pennines in Cheshire and the Staffordshire coalfield. It formed when Irish Sea ice impinged on the west Pennine scarp in the Devensian to produce a sweeping belt of hummocky topography. Other significant areas of supraglacial deposition occurred in the Vale of York and in the vicinity of the Solway Firth. Sheets of outwash gravels are relatively less extensive in the British Isles than in the great north European lowland, though notable exceptions occur, as in north Norfolk or the Severn lowlands.

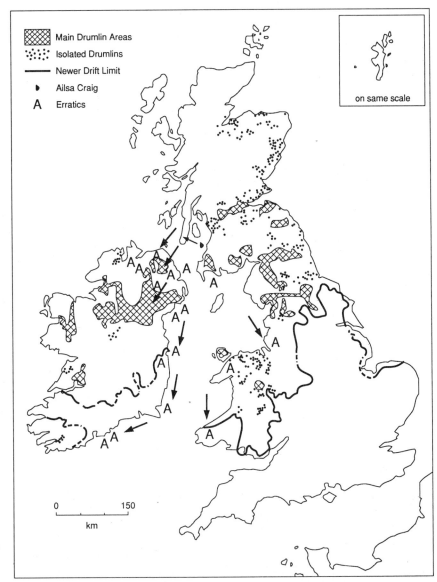

Fig. 26. The distribution of drumlins and erratics derived from the outcrop of Riebeckite-Eurite on Ailsa Craig.

The periglacial impact

It can be argued that some parts of the British Isles have a periglacial environment today. High altitude areas like the Scottish Highlands or Snowdonia are subjected to the effects of intense and frequent frosts, and of long continued and substantial snowfalls. It is, therefore, not surprising that such areas have patterned ground produced by frost effects, active solifluction lobes, some aeolian deposits, and snow avalanches.

During the glacial phases of the Pleistocene, periglacial influences were far more extreme, and large tracts of the British Isles were subjected to conditions of permanently frozen ground (permafrost). Because of the moderating influence of oceanic conditions, the far south-west of Britain and Ireland may have escaped some of the most extreme effects of periglaciation such as permafrost, but even there it is possible to find manifestations of the power of periglacial processes at low altitudes.

There are several relatively clear geographical patterns in the distribution of periglacial landforms (Figure 27). The first is that the greatest frequency of such forms is concentrated in the southern portions of the British Isles. In particular, the largest concentrations occur to the south of the limit of glaciation during the last glacial advance (the Devensian of Britain and the Midlandian of Ireland). This can be explained by the fact that, because they were not covered by the ice, such areas would be exposed to the full rigours of a tundra

environment. Indeed, given that permafrost requires negative mean annual temperatures, and that widespread permafrost requires a mean annual temperature of less than about –4 °C, temperatures at the time that permafrost was present may have been as much as 14 °C lower than those experienced today.

The second pattern that deserves mention is that certain rock types appear to display the effects of periglacial processes rather more clearly than others. This is especially true of the chalk landscapes of southeast England. Chalk is a rock that, because of its physical properties, is highly susceptible to frost attack, while its high permeability means that the periglacial forms have not been erased by Holocene fluvial geomorphological processes to the extent that they may have been on some other rock types.

A third pattern that can be identified is the influence of the short-lived Loch Lomond re-advance that occurred round about 11 000 years ago. Glaciers reappeared in various parts of Highland Britain, and just down-valley from the limits of the Loch Lomond glaciers there are often suites of very clear periglacial features, including screes, protalus ramparts, and the like.

The map is a little confusing because periglacial features do occur within the *maximum* ice limits. Obviously many features were formed during deglaciation. The ring of symbols in eastern Scotland clearly reflects both a decaying ice sheet and the Loch Lomond pattern.

Fig. 27. Distribution of late-Devensian periglacial structures.
(*Source:* Rose, fig. 18.2 in Richards *et al.* 1985.)

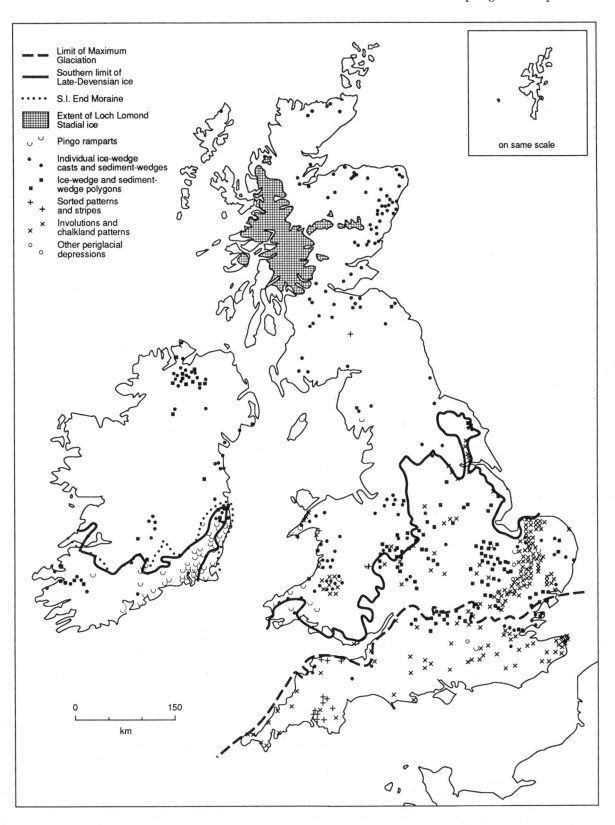

Limit of Maximum
Glaciation

Southern limit of
Late-Devensian ice

S.I. End Moraine

Extent of Loch Lomond
Stadial ice

Pingo ramparts

Individual ice-wedge
casts and sediment-wedges

Ice-wedge and sediment-
wedge polygons

Sorted patterns
and stripes

Involutions and
chalkland patterns

Other periglacial
depressions

on same scale

0   150
km

Aeolian materials

During the periods of glacial activity aeolian action was very important. The coastal sediments were widely exposed and provided abundant sources of fine material. The tundra environment was hostile and the landscape was not heavily vegetated. In addition fine material was easily deflated from the outwash and glacial deposits immediately in front of the ice sheets. Thus there are extensive spreads of aeolian material, particularly in southern Britain.

The most widespread of the inland materials are deposits of windblown silt, called loess (Figure 28). In comparison with the very extensive deposits found in neighbouring parts of Europe (e.g. in the Rhine Valley, the Paris Basin, and Normandy) it is primarily patchy and thin in character. Often it has been mixed into underlying materials by periglacial frost churning, and it has been removed from many areas by such processes as solifluction, stream erosion, and accelerated soil erosion resulting from deforestation and cultivation during the Holocene. Valley fills contain thick layers of secondary loess. Primary loess is best preserved on calcareous rocks, possibly because of enhanced cementation as a result of the close proximity of lime, but more probably because it is less easily removed by water erosion from permeable surfaces that occur on limestone and dolomite.

The thickest deposits of loess, which reach a maximum thickness of 4 m at Pegwell Bay in Kent, are in south-east England, where they form a continuation of the more extensive European deposits. Relatively little loess occurs to the north of the limit of Devensian glaciation largely because loess was probably formed under a rigorous periglacial tundra landscape as a result of deflation from glacial outwash and moraine. Most deposits appear to have a late Devensian age.

Another major type of aeolian deposit of Pleistocene age is coversand, which results from the degradation of ancient dunes. Such ancient dunes are a feature of many dry periglacial landscapes, and are very extensive in areas like southern Sweden. Aeolian sands occupy the southern parts of the Vale of York and the lower Trent valley near Scunthorpe. Other late Devensian coversands occur in south-west Lancashire (the Shirdley Hill Sands), and in the Breckland of East Anglia. Small patches also occur banked up against the Cotswold scarp near Cheltenham, and were probably derived from wind-winnowing of Severn Valley outwash plains in late Devensian times. Some aeolian activity occurs in the highlands of Britain at the present time. Vegetation-covered 'sand sheets' occur, for example, on the Torridon Sandstone massif of An Teallach in north-west Scotland. Indeed, such forms are best developed in sandstone and granite terrains.

An aeolian curiosity is the occurrence of ventifacts—wind-sculpted pieces of rock more commonly associated with wind-swept desert surfaces. Examples are found in the Midlands and near Chelford on the Cheshire Plain.

Fig. 28. The distribution of vegetation-covered sand sheets (over 200 m) in elevation, coversands, and loess.

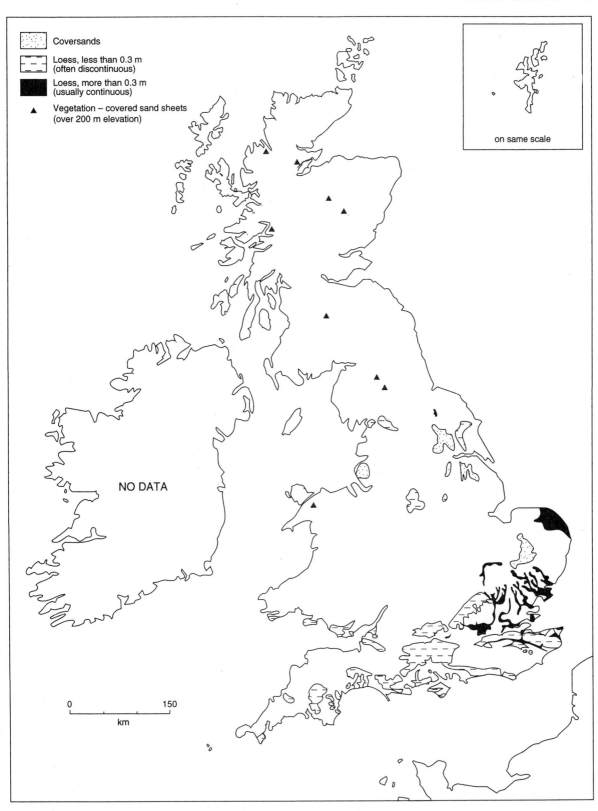

Coversands

Loess, less than 0.3 m
(often discontinuous)

Loess, more than 0.3 m
(usually continuous)

▲ Vegetation – covered sand sheets
(over 200 m elevation)

on same scale

NO DATA

0   150
        km

Limestone landscapes

Limestones of great diversity outcrop over the British Isles (Figure 29). They include the chalk of the Cretaceous (which outcrops in southern and eastern England, and in Antrim), the oolites of the Jurassic (which extend up through 'the backbone of England' from Dorset to Yorkshire), the magnesium-rich limestones and dolomites of the Permian, the Devonian limestones (which outcrop in south Devon), the Carboniferous limestones, and the ancient Ordovician limestones of north-west Scotland. Even clay strata include thin limestone bands or calcareous mudstone.

Karstic features (e.g. caves, collapse hollows, swallow holes, dry gorges, etc.) tend to be best developed on the older limestones (i.e. those of the Ordovician, Devonian, and Carboniferous) because they tend to be less porous and permeable than later limestones, and have greater overall strength. Of all these formations it is the Carboniferous Limestone that is associated with the largest areas of well-developed karst, and splendid examples occur in the Burren of County Clare and other parts of Ireland, north-west Yorkshire, the Peak District of Derbyshire, the Morecambe Bay area, the Mendips, the south Wales coalfield area, and portions of north Wales. None the less, the importance of karst on other formations should not be dismissed. For example, large numbers of closed depressions occur on parts of the Jurassic Lincolnshire limestone, and the chalklands of Dorset are pock-marked by collapse depressions such as Culpepper's Dish, especially where there are strong hydraulic gradients to dry valleys and the chalk has been dissolved by acidic waters draining from Tertiary sands and gravels,.

Besides the control on landform distribution exercised by rock type, it is also necessary to bear in mind the imprint of glacial erosion in Pleistocene times. Thus, the major expanses of limestone pavement, such as those of the Burren, the Ingleborough area, and areas near Morecambe Bay, occur in zones that were severely glaciated. The pavements were revealed by glacial stripping of the overlying soils. To the south of the glacial limit, as for example in the Mendips, although there are areas of suitable Carboniferous limestone upon which pavements could have developed, no pavements exist because the surfaces were never subjected to glacial attack.

Although limestone solution is the prime cause of specific limestone landforms there are many examples in the British Isles where small landforms have been created by the deposition of calcium carbonates in the form of tufa. This is a soft, porous rock that forms in springs, waterfalls, and lakes as a result of chemical changes and biological influence. The landforms produced include tufa dams, barrages, mounds, and waterfall curtains. Excellent examples occur in the Malham and Goredale regions of north-west Yorkshire, and in the Caerwys area of north Wales.

Another neglected feature is the distribution of ancient karst features. There is geological evidence of palaeo-karst activity such as the intra-Carboniferous karst and the sub-Triassic of South Glamorgan. These episodes do not, however, really affect British landforms today. More important are palaeo-solution depressions and half-emergent limestone pinnacles. Such depressions filled with Blackheath Bed gravels lie just beneath the flat surface of the North Downs at Sanderstead or beneath and alongside the trunk road at Kingsteignton in Devon. Here they are filled with Upper Greensand, sands, clays, and lignite and may even represent the equivalent of the tropical karst found on the other side of the Cretaceous sea in Bavaria!

Fig. 29. Areas of calcareous rocks and tufa deposition sites in Britain together with known rates of solution.

(*Source*: Pentecost 1978.)

1	Pentland Hills	8	Howgill	15	Caerwys	21	Dover
2	Helmsley	9	Aysgarth	16	Bourne Brook	22	Helen River
3	Gordale Beck	10	Knaresborough	17	Durley	23	Clayton
4	Malham Tarn	11	Matlock Bath	18	Harpenden	24	Lewes
5	Waterfall Beck	12	Penrhyn Glas	19	Plaxtol	25	Totland Bay
6	Kettlewell	13	Penmon	20	Wateringbury	26	Blastenwell
7	Clapham Beck	14	Menai Straits				

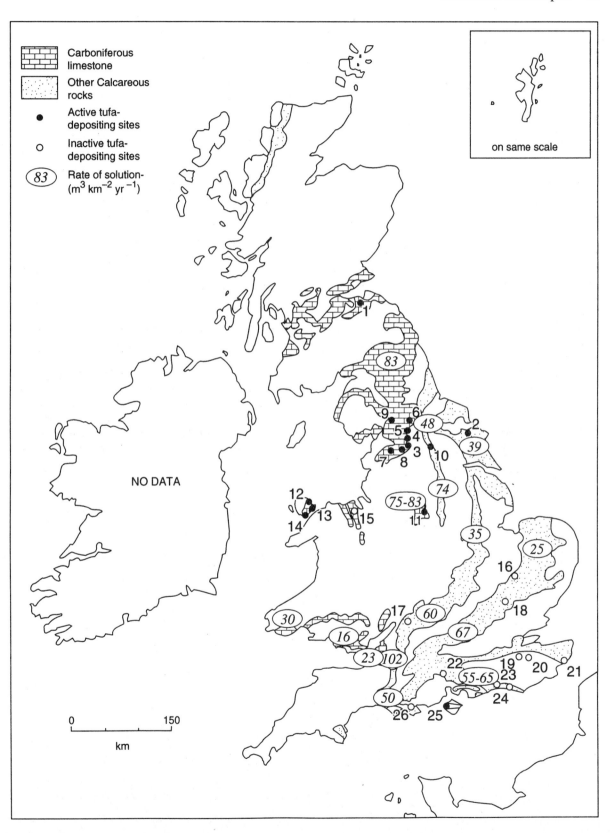

Landslides

There are many thousands of landslides in Britain including falls, rotational and translational slides, flows, toppling failures, and many complex varieties. Although Britain is a relatively stable area tectonically it possesses many of the other attributes that favour landsliding. The coast is under constant attack from the sea. The sea-level is rising and therefore erosion is increasing. In the past sea-level has fluctuated causing river incision and steep slopes. The climate delivers enough rain and severe storms to promote high water pressures and slope failure. Climatic change means that there have been wetter conditions and more landsliding in the past, and glaciation has created many steep and unstable slopes. Ground freezing, solifluction, and cambering have all been associated with landsliding.

Geological factors are fundamental controls. The scarpland and the coasts of southern England are prime sites because here the simple geological structures and the presence of permeable chalks and sands overlying clay sequences provide classic landslide conditions. The main vulnerable rock types are clays (22% of total number of slides), shales (11%), sandstone (9.5%), and limestone and interbedded argillaceous materials (7%). Siltstones, mudstone, and schists (5%) are also dangerous.

The main causes are erosion, which steepens and heightens the slope (69%), change to the water conditions (13%), and weathering (8%). Human activity is also a common cause of failure. Rotational slides (12%), planar slides (11%), flows (5%), and complex varieties (11%) are the common forms.

The main areas are the Weald, London Basin, Cotswolds, southern Pennines, south Wales, the Welsh border, and the east Midlands plateau (Figure 30). The most affected coastal areas are Gwynedd, Lyme Bay, Purbeck, Isle of Wight, north Kent and Sandwich, Hastings and Hythe, the Cromer to Overstrand coast, and Robin Hood's Bay. In Scotland debris flows in the Highlands occasion major rockfalls from the glaciated slopes. The Storr on Skye forms one of Europe's major landslide complexes.

Most recorded landslide impacts on the community of course occur in populated areas such as London, Avon, Gwent, and Glamorgan. Seventy-nine per cent affect road, rail, and community property. Some towns such as Ventnor or Shanklin are vulnerable but adopt very sensible management policies. A very large problem is the reactivation of ancient slopes by construction because the past record of climatic change means that there are many relict landslide features. Despite this, however, most landslides in Britain are small, slow, and cause a low loss of life. There are rare disasters but it is unusual for there to be widespread damage.

Fig. 30. Landslides per 100 km².

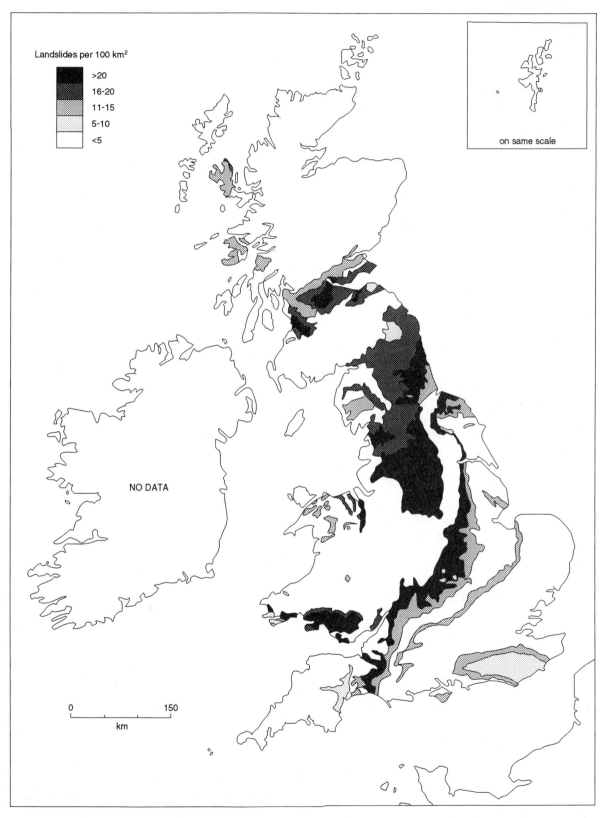

Landslides per 100 km²

>20
16-20
11-15
5-10
<5

on same scale

NO DATA

0   150
km

Coastal dunes

Coastal dunes are an important landform type and habitat around the coast of the British Isles. In England and Wales they constitute about 9% of the total coastline, and extend over 400 km of coast length. In Great Britain as a whole they occupy an area of over 56 000 ha. In Ireland dunes are relatively even more extensive, extending over 1000 km, comprising about 20% of the coastline and covering an area of 160 km².

Although sand dunes are found along many parts of the coastline, there is a marked geographical bias towards the western coast of Ireland, the Western Isles of Scotland, Orkney and east Scotland, Wales, the northern coast of the south-west peninsula of England, and north-east England (Figure 31). Dunes are relatively rare on the south coast of England, though there are some notable exceptions to this generalization. The reasons for the paucity of dunes along the south coast are not entirely clear, but large stretches of the coastline are backed by cliffs and so do not have suitable sites for dune accumulation. Neatly compartmentalized bays of the type found along the west coast, and behind which dunes may form, are here less well-developed. Moreover, the tidal range is often low, relative to the beach and onshore gradient, thus giving narrower intertidal zones from which deflation of sand can occur to supply dunes. In addition, the sea floor may have received less suitable dune-forming material as a result of the lesser input of glacial material to the continental shelf than in more northern areas. Many south coast beaches are composed of cobbles and shingle with the intertidal area containing a high silt fraction.

The dunes of the British coastline also show considerable variability in the materials of which they are composed, but certain geographical patterns can be identified. In particular the dunes of western Ireland, northwestern Scotland, Pembroke, Devon, and Cornwall often have a high carbonate content (12–87%). In some parts of Cornwall the carbonate content may amount to over 70% (Constantine Bay), whereas some of the dunes on the south coast of England may have carbonate contents of less than 1% (Studland).

There appear to be five main types of coastal dune in the British Isles. *Offshore island dunes* are those developed on offshore (barrier) islands and serve to protect the mudflats in their lee. Good examples occur on the north Norfolk coast at Blakeney. *Prograding ness dunes* form on an open coast where there is an abundant sand supply at an accumulating point (ness) such as Winterton. *Spit dunes* form on sandy promontories at the mouths of estuaries, and often form a fan-like series of ridges with intervening depressions (slacks). Studland is of this type. *Bay dunes* occur in bays such as Hope Cove, Devon, along indented coastlines, while *hindshore dunes* occur on extensive sandy coasts where the prevailing wind is also the dominant one. Large dunes travel inland as great arcs or ridges (e.g. at Newborough Warren in Anglesey).

Many dune systems, perhaps the majority, display marked erosion and relatively few are showing accretion. From the *machairs* of the north and west of Scotland, to the dunes of Camber sands in the far south-east of England, dunes are being subjected to considerable pressures from human overuse and misuse. This is a matter of concern because dunes are both valuable natural habitats and perform an important sea-defence function.

Fig. 31. Major dune areas together with an indication of their main chemical type. (*Sources*: Goudie 1990, fig. 8.10; Doody 1985, fig. 1; and Carter 1988, fig. 1.)

Strathy Bay Dunes **C**
Invernaver   **C**
Faraid Head & Bainkeil Bay Dunes **C**
Sandwood Bay Dunes   **C**
Northton Bay Machair   **C**
Berneray Machair   **C**
Baleshare & Kirkibost Machair   **C**
Monach Isles   **C**
Loch Bee Machair   **Ca**
Howmore Estuary Machair   **C**
South Uist Machair   **C**
Crossapol & Gunna Machair   **C**
Hough Bay & Ballevullin Machair   **C**
Killinallan Dunes   **C**
Laggan Bay Dunes   **C**

Morrich More   **A**
Culbin Sands
Loch of Strathbeg Dunes   **C**
Sands of Forvie   **A**
St. Cyrus   **C**
A
Tentsmuir & Earlshall Muir   **Ca**
Lindisfarne (Holy Island Dunes & Ross Links)   **Ca**

Irvine Bay

Beckfoot

Torrs Warren Dune   **A**

Drigg Point Dunes   **Ca**
North Walney & Sandscale Dunes   **Ca**

Spurn Head   **C**

Sefton Coast Dunes   **C**
Saltfleetby & Theddlethorpe Dunes   **C**
Gibraltar Point Dunes   **C**
Prestatyn

Tywyn Aberffraw   **C**
Newborough Warren   **C**
Morfa Harlech Dunes   **C**
Ynyslas Dunes   **C**

North Norfolk Dunes   **C**
Winterton Dunes   **A**

Morfa Dyffryn   **C**
Barmouth   **C**

Whiteford Burrows   **C**
Kenfig Dunes   **C**
Merthyr Mawr

Whitesand Bay

Sandwich & Pegwell Bay Dunes   **C**

Stackpole Warren   **C**
Tywyn Gwendraeth Dunes   **C**
Oxwich Dunes   **C**
Weston-s-Mare
Brean
Camber
Braunton Burrows   **C**

Coastal dunes, *(together with south Wales above)*, containing +20% CaCO₃
Bude

Littlehampton
Studland Heath Dunes   **A**

Dawlish Warren   **A**

● Dunes

CHEMICAL TYPES
C   Calcareous
Ca   Calcareous/Acidic
A   Acidic

AREA OF DUNE SYSTEMS (ha)
8000
4000
1000
100

0   150
km

on same scale

Coastal erosion

The rate of cliff recession in parts of the east and south coasts are among the fastest measured rates in Europe. The highest rates can exceed an average of 1 m per year at Holderness, Pakefield (Norfolk), and between Seaford Head and Hastings on the south coast (Table 2). Coastal agricultural land, cliff-top houses, local roads, and even whole villages have been lost to the sea. This loss is so serious that over 11.5% of the British coast is now protected by sea defences.

This situation is caused by the power and range of the tidal streams, the height and period of the abundant wave energy, the rising sea-level, and the occurrence of easily erodible coastal materials. In west Dorset, for example, Jurassic clays and sands form unstable cliffs over 150 m high and erode at 0.5–1.0 m per year. In Holderness unconsolidated glacial materials have very little resistance. At Pakefield in Norfolk glacial drifts have disappeared at the rate of 2.94 m a year over the last 64 years.

The highest rates occur where these factors coincide: Bridlington, Spurn Head, Sheringham to Happisburgh, Lowestoft and Southwold, the Naze, Herne Bay, Sheerness, Selsey Hill, Isle of Wight, Hurst Castle, west Dorset, the Dee, and St Bees Head.

Solutions range from the enhancement of natural defence systems such as beaches and dunes to sediment-control structures, groynes, strong points, and then sea walls and breakwaters (Figure 33). Increasingly it is being realized that the choice rests in an understanding of the natural systems. In the past many problems have been caused by protection works being designed within planning-authority boundaries and not natural process cells: protection in one place causing erosion in another because the system was not known. It is welcome news that coastal authorities are now grouped into units that consider the natural conditions as a whole. The planning units are known as coastal defence groups (Figure 32).

Table 2. Rates of cliff erosion

Area	Geology of cliff	Average rate of retreat (m/100 y^{-1})
North Yorkshire	Shale	9
North Yorkshire	Glacial drift	28
Holderness	Glacial drift	120
Norfolk		
Weybourne–Cromer	Glacial drift	42
Cromer–Mundesley	Glacial drift	96
Mundesley–Happisburgh	Glacial drift	88
Gratby–Caister	Glacial drift	83
Gorleston–Corton	Glacial drift	57
Pakefield–Kessingland	Glacial drift	105
The Naze (Essex)	Glacial drift, London Clay and Crag	11–88
Kent		
Reculver	London Clay	68
N. Isle of Sheppey	London Clay	96
Isle of Thanet	Chalk	7–22
St Margaret's Bay–Folkestone	Chalk	7–19
Folkestone	Gault Clay	28
East Sussex		
Peachhaven	Chalk	46
Seaford Head	Chalk	126
Birling Gap	Chalk	122
Beachy Head	Chalk	106
Ecclesbourne Glen	Hastings Beds (sandstone)	119
Fairlight Glen	Hastings Beds (clays)	143
Cliff End	Hastings Beds (sandstone)	108
Hampshire		
Christchurch Bay (Highcliffe Castle)	Bracklesham Beds	3
Christchurch Bay (Barton)	Barton Beds	58
Christchurch Bay (Hordle)	Headon Beds	18
Dorset		
Ballard Down	Chalk	23
Kimmeridge Bay	Kimmeridge Clay	39
White Nothe–Hambury Tout	Chalk	21
Ringstead	Kimmeridge Clay	41
Furzy Cliff–Short Lake	Kimmeridge Clay	37
Isle of Wight		
Cranmore	Hamstead Beds	61
Newton River–Gurnard	Bembridge Beds	38
Brighstone Bay	Wealden Beds	52

Fig. 32. Coastal defence groups around England and Wales.

Fig. 33. The distribution of erosion rates, artificial embankments, and groynes around the coastline of England and Wales.

Coastal cliffs

Because of the combination of suitable rock types and the high energy location, large stretches of the coastline of Britain consist of high cliffs. Such cliffs, which often exceed 30 m in height, occur on coasts of all aspects, and face the English Channel, the North Sea, the Irish Sea, and the North Atlantic.

Cliffs also occur on a wide range of rock types, from the glacial debris of north Norfolk (near Mundesley) to the chalk of Flamborough Head and Beachy Head, the Purbeck and Portland limestones of the Isle of Portland (Dorset), the sandstones of St Bees Head, the Old Red Sandstone and Carboniferous Limestone of Gower and Pembroke, the granite masses of Land's End, the Tertiary lavas of Antrim, and the ancient rocks of north-western Scotland (Figure 34).

Steep vertical cliffs are favoured where horizontal or gently dipping sedimentary rocks are oversteepened by wave attack, where highly erodible materials (like glacial drift) are being rapidly attacked by the sea, or where resistant rock masses act as bastions against the attack of powerful breakers.

Some parts of the British coast display the effects of major mass movements, and these contribute a great deal to cliff morphology. In particular, major slips occur in susceptible rocks, especially the clays and shales of southern and eastern England. These slips can take the form of deep-seated rotational landslides (for example, Folkestone Warren in Kent, Stonebarrow in Dorset, and the undercliffs of parts of the Isle of Wight), or long, shallow, translational mudslides, such as those that have developed in the clays that border the Thames Estuary.

There is also a geographical pattern in the rate at which cliff erosion takes place, with the highest rates occurring in the unconsolidated Pleistocene deposits of the east coast, or in such susceptible rocks as chalks and shales (Figure 33). In exposed locations, such materials may erode at rates as high as 100 m per century. In contrast, many of the cliffs of the northern and western coasts of the British Isles, developed in resistant, ancient rocks, have a very long history, have witnessed many changes of climate and sea-level, and, in some cases, are now being exhumed from beneath Pleistocene glacial drift or soliflucted material.

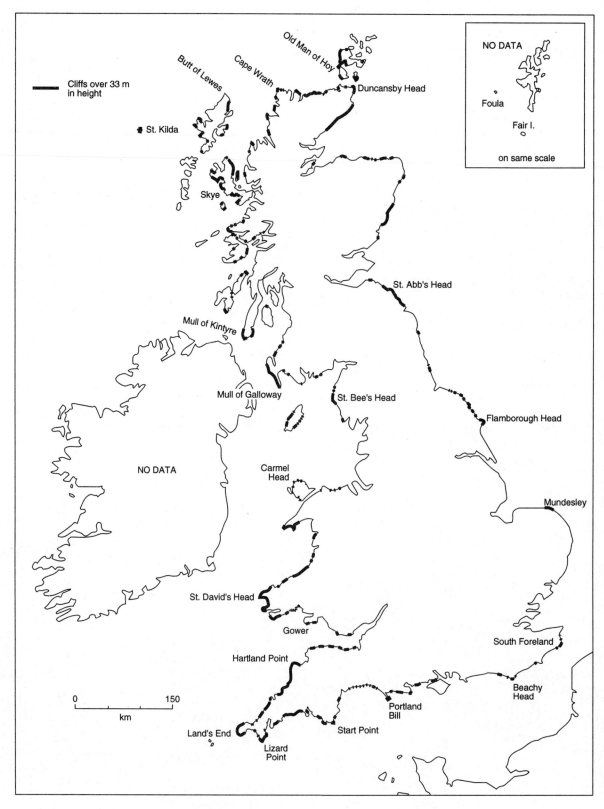

Fig. 34. The location of major cliffs (Great Britain only).

Salt marshes

The flora of salt marshes show some geographical variation in the British Isles. Paul Adam (1990) suggests that there are three main groupings (Figure 35). Type A marshes are characterized by a relatively low community diversity and a flora comprised largely of salt-loving (halophytic) plants such as *Salicornia* (marsh samphire). *Spartina* is often widespread, as is *Aster tripolium* (sea aster). Creeks are often lined by *Halimione portulacoides* (sea purslane). Such Type A marshes occur mostly in south-east England. Type B marshes may have widespread *Spartina*, but they are dominated in the lower areas by *Puccinellia maritima* (salt-marsh grass). Their upper and mid-marsh zones are predominantly grassland, with much *Juncus*, whereas the marsh fringe in the upper parts of Type A marshes is often occupied by *Elymus pycnanthus* (*Agropyron pungens*). The Type B marshes are extensive in north-west England and Wales. Type C marshes have relatively few communities but are occupied by fen-like communities in their upper portions. Salt-loving species are relatively rare. Most marshes of this type are located in western Scotland.

The differences between the Type A and Type B marshes has been explained by two main mechanisms: the nature of the substrate and the degree of grazing pressure by domestic animals. The marshes of the north-west and Wales are more heavily grazed than most of those in south-east England, and also tend to have a sandy substratum rather than the silts and clays of the south-east. It is possible that the two mechanisms are related in that grazing is favoured on the sandier marshes. The Type C marshes of western Scotland are grazed, but another important determinant of their character is the severe climatic conditions which they have to endure compared to those further south. Some species are absent for that reason. Moreover, the high regional rainfall, and the influence of seepage from the land reduce the halophytic element.

A very important influence on the composition and form of British marshes has been the spread of a form of cord grass, which was first collected from Southampton Water in 1870 (Figure 35). This cord grass, called *Spartina townsendii* (if sterile) and *Spartina anglica* (if fertile), is probably the result of natural hybridization between a native form, *Spartina maritima*, and an American form that may have come in with ships' ballast, called *Spartina alterniflora*. The new hybrid has spread explosively, and although there has been some die-back in areas like Poole Harbour, it is now very common all round the coast from the Isle of Harris to the Cromarty Firth. It has been very effective at causing marsh accretion by trapping sediment, and has largely replaced some native communities such as *Zostera* sea-grasses. Rates of accretion can be as high as 8–10 cm y^{-1} (Bridgwater Bay) to 2 cm y^{-1} in Poole Harbour. The Dee and the Severn also show enhanced rates. These rates should be compared with 1.7 cm y^{-1} on 10-year-old marshes in north Norfolk to 0.002 cm y^{-1} on 500-year-old marshes at Scolt Head. In the Severn there has been 1.2 m of accretion in the last 1900 years but 0.21 m since 1945 due to *Spartina*.

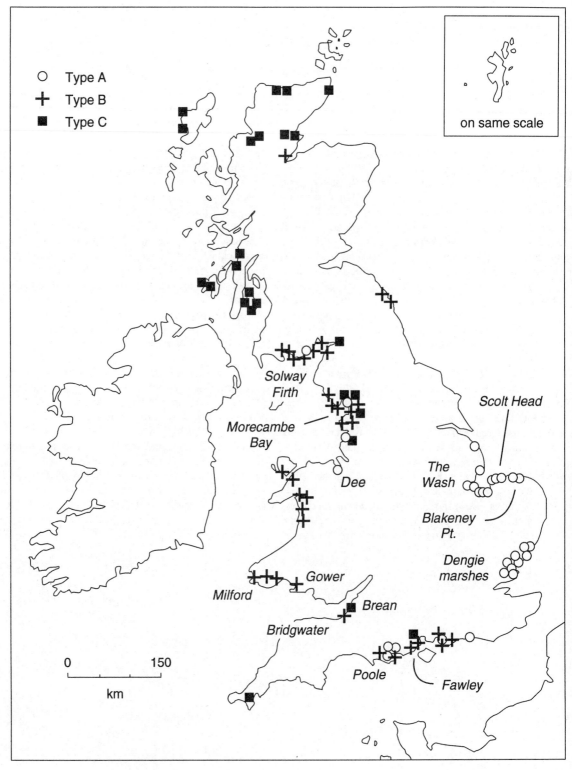

Fig. 35. The distribution of Type A, B, and C salt marshes (Great Britain only).
(*Source*: Adam 1990, fig. 3.7*b*, with modifications.)

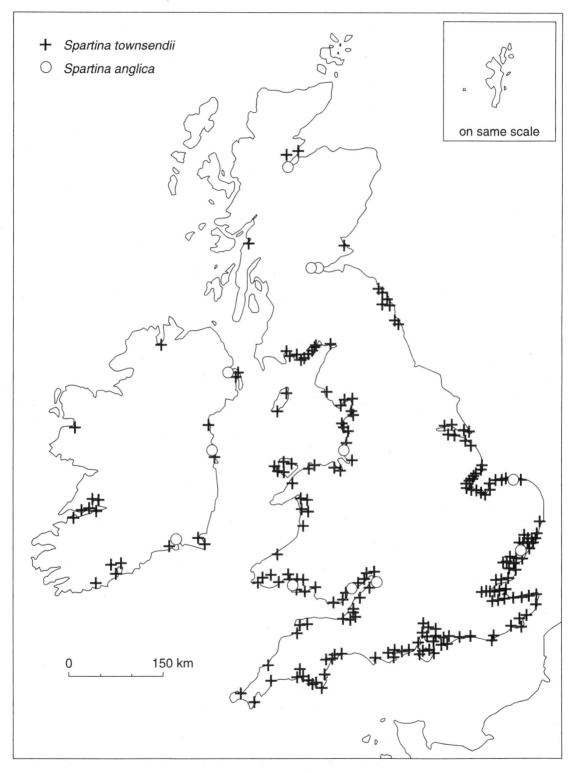

Fig. 36. The current distribution of *Spartina townsendii* and *Spartina angelica*. (*Source*: Adam 1990, fig. 2.1.)

Spits

The British coast is famous for its examples of spits and other coastal constructional forms. Many are made of shingle, usually remnant deposits brought to the coast during the post-glacial rise of sea-level and therefore thought to be unusual because they have no obvious source of supply today. Slapton Beach made of flint and chert is a good example. Chesil Beach, really a tombola and barrier beach, is another fine example containing, as it does, pebbles from as far down the coast as Budleigh Salterton. These could only have reached this far along the coast at a low sea-level. An implication of this is that many British spits are now cut off from their original offshore supply and unless they are supplied by erosion of cliffs along the coast will diminish in size. There is also a clear tide- and wave-process control over their distribution (Figure 37). In coastal areas where tidal ranges are over 4 m tidal sand flats and saltmarshes are common. If, however, tidal range is less than 2 m the main process is wave energy and longshore drift. Here beaches, spits, barrier beaches, and cuspate forelands develop.

These areas also lie 'down drift' from eroding cliffs and sediment supply zones. Therefore at the smaller scale attention focuses on the sediment or process cell. Lyme Bay with the landslides feeding Chesil, Christchurch Bay feeding Hurst Castle, Yarmouth feeding Orford Ness, or Holderness feeding Spurn Head are examples.

Supply also comes from the estuary across which the movement takes place. Some spits and barriers grow efficiently relative to the river deposition and manage to close the mouth. Loe Pool in Cornwall, Slapton Beach, Charmouth, or Burton Bradstock have achieved this. In rivers like the Humber and Mersey, however, this is not possible. Instead the spit itself often breaches and the resulting island is driven onshore to infill the estuary.

Occasionally, the shingle structure adopts a cuspate shape such as Dungeness. This usually occurs by complex breaching of sediment structures across a wide embayment where river movement is also important. Human interference should always be considered because spits are annoyances if they grow too much, worrisome if they breach or erode, and yet regarded as a resource to be mined as required!

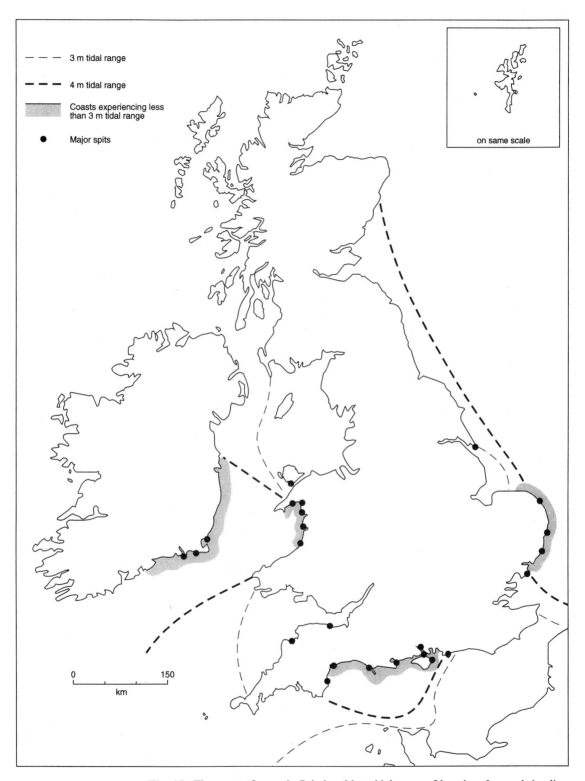

Fig. 37. The areas of coast in Britain with a tidal range of less than 3 m and the distribution of major spits.

(*Source*: Pethick 1984, fig. 4.17; and Goudie 1990, fig. 8.8).

Areas at risk from sea-level rise

If, as a result of the greenhouse effect, caused by the increasing levels of certain gases in the atmosphere (notably carbon dioxide, methane, nitrous oxide, and chlorofluorocarbons), sea-level rises, then certain areas of the British Isles might be at risk from coastal flooding.

There are still considerable uncertainties about the degree of sea-level rise that may be anticipated in future decades, and published predictions for the year 2100 vary between about 0.5 m and 3.5 m. The Intergovernmental Panel on Climate Change, which reported in 1990, suggested that the lower sorts of estimate were the most likely, and indicated that the mean rise would be about 65 cm by 2100. This rate would be exceeded in areas that were being subjected to subsidence, such as parts of south-east England. It is also possible that the height and frequency of flooding will be affected by climatic changes. Storms might, for example, become more frequent and intense and, because the tidal pattern and deposition areas will change, coastal flooding and erosion may change location.

Many people and many human activities are located in low-lying areas, so that if may be anticipated that, if sea-level rises even by quite modest amounts, the consequences might not be entirely trivial. Many of the areas at potential risk are conurbations (e.g. parts of London), high-grade agricultural land (e.g. the Fens), or major industrial installations (e.g. power-stations and oil refineries).

Figure 38 shows those areas of Britain which lie less than 5 m above present sea-level. Although projected sea-level rises are at present much less than this figure, tidal surges and sea-level rise combined could have adverse effects on some of these zones (Figure 39).

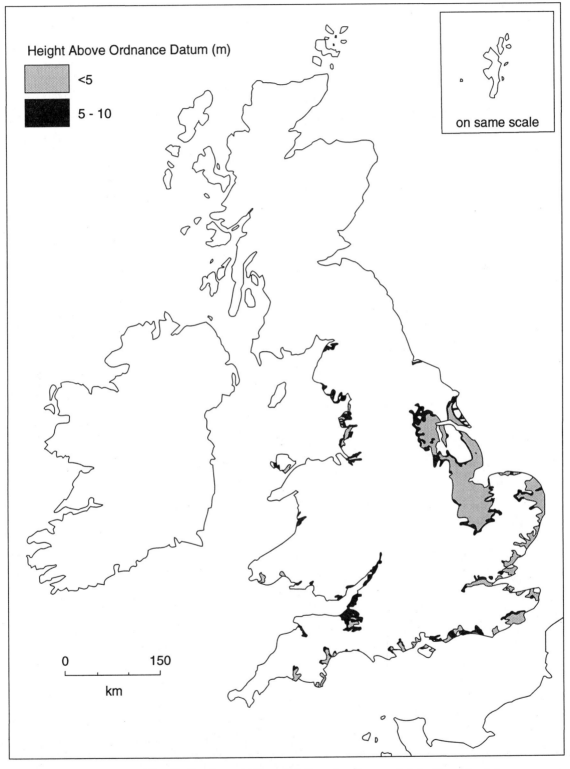

Fig. 38. Land potentially at risk from flooding associated with sea-level rise.
(*Source*: Whittle 1989.)

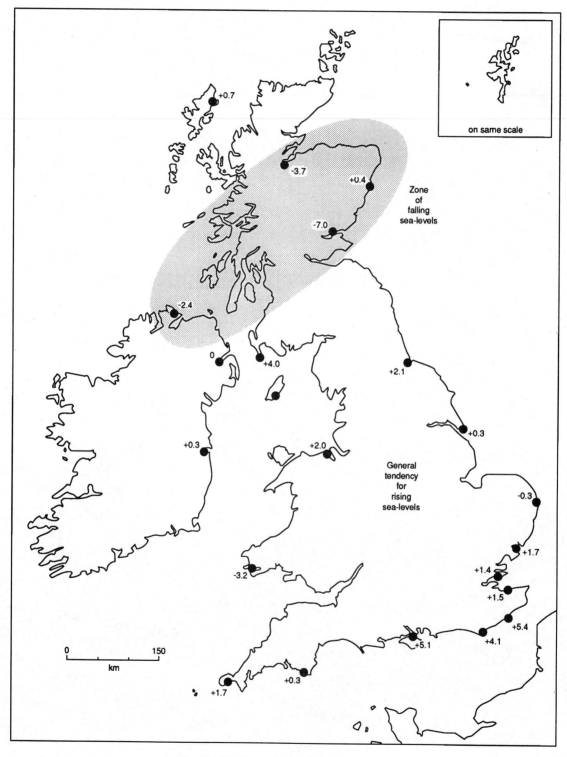

Fig. 39. Recent sea-level changes in mm per year.

(From Carter 1989.)

Climate

Annual rainfall and its variability

The rainfall of the British Isles is much measured and discussed, and for such a small area there is a great range of values between different parts of the country. Near Sprinkling Tarn in the Lake District the mean annual total is around 5000 mm, whereas at the other extreme, in the English Fenland and the Thames Estuary, totals may be as low as about 500 mm—a tenfold range (Figure 40).

Annual totals are high in the higher areas of Ireland, Wales, Scotland, and the English Lake District, and gradually decrease to the south and east. The first explanation for this pattern is that most precipitation comes from frontal depressions that move in from the Atlantic. They reach their maximum frequency and intensity in the area most often crossed by the fronts that border the Icelandic Low. This is located to the north-west of the British Isles. The second reason for the rainfall distribution is that the presence of the highest mountains in the north and west serves to exaggerate this pattern because of orographic amplification. All the mountainous areas of the British Isles show the orographic effect from Dartmoor with >1000 mm to the Brecon Beacons and Snowdonia with 2400 mm, the Pennines with 1600 mm, and Cumbria and the Highlands with over 2400 mm.

Although the British Isles are blessed with relatively reliable rainfall in comparison with many parts of the world, such as the arid and semi-arid areas, there is some variation from year to year in response to the location, intensity, and duration of the cyclones and blocking anticyclones that control so much of the character of the climate of the area.

One method of expressing inter-annual rainfall variability is to calculate the coefficient of variation, which relates the standard deviation of the data to the average, expressed as a percentage (Figure 41). The map therefore shows how variable the rainfall is in the sense of how different it can be from the most commonly experienced condition. The greatest variability occurs in the south-east of England (averaging around 18%), and the lowest variability in north-western Ireland and north-western Scotland (averaging around 10%). For comparison, values in the Sahara range from 80 to 150%!

The increased variability in the south is due to the greater likelihood of convectional storms which can deliver a third of the annual rainfall in a single event. For example, summer storms are very common in the south-east and may cause disastrous flooding because the rainfall intensities are so high (Figure 42). The average picture is revealed by the contrast between two fascinating maps. Figure 42 portrays the greatest rainfall produced in one hour by storms occurring during any 5-year period at any place. Thus in parts of central and eastern England over 20 mm might be expected to fall in one hour once in five years. In comparison the distribution for the rainfall to be expected in 48 hours (Figure 43) shows that the high areas to the west expect rainfalls of over 75 mm in 48 hours once in every five years. The one-hour rainfalls reflect the dominance of concentrated, convective storms. The two-day totals are obviously provided by westerly systems and orographic effects. Figure 43 could, therefore, almost be regarded as a replica of a relief map.

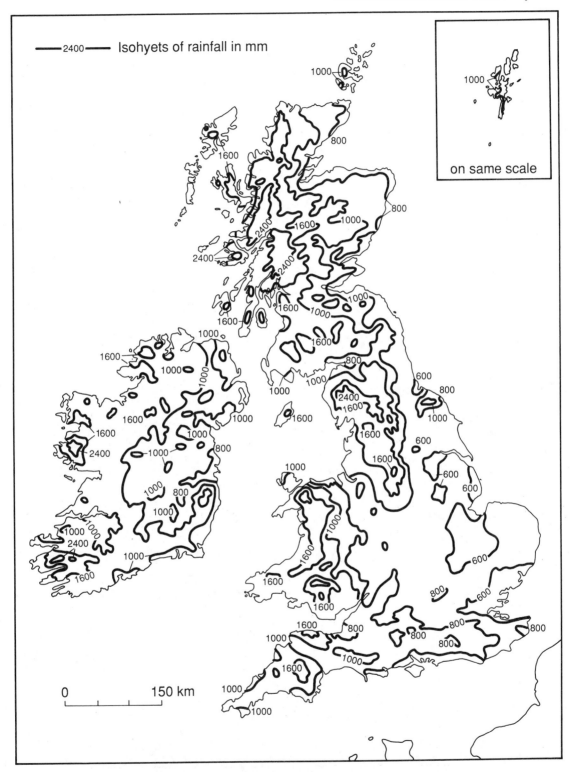

Fig. 40. Mean annual rainfall in mm, for the period 1931–60.

(Based on data provided by the UK Meteorological Office and the Eire Meteorological Service.)

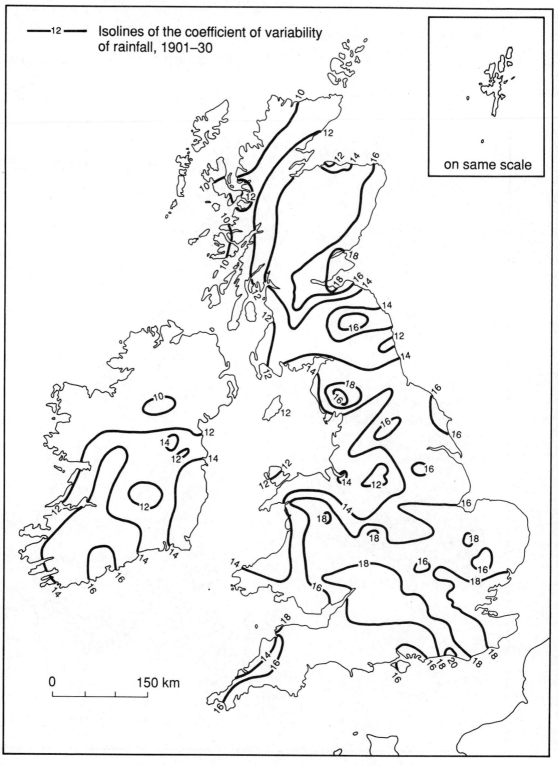

Fig. 41. The coefficient of variability of annual rainfall, for the period 1901–30. (From Gregory 1955.)

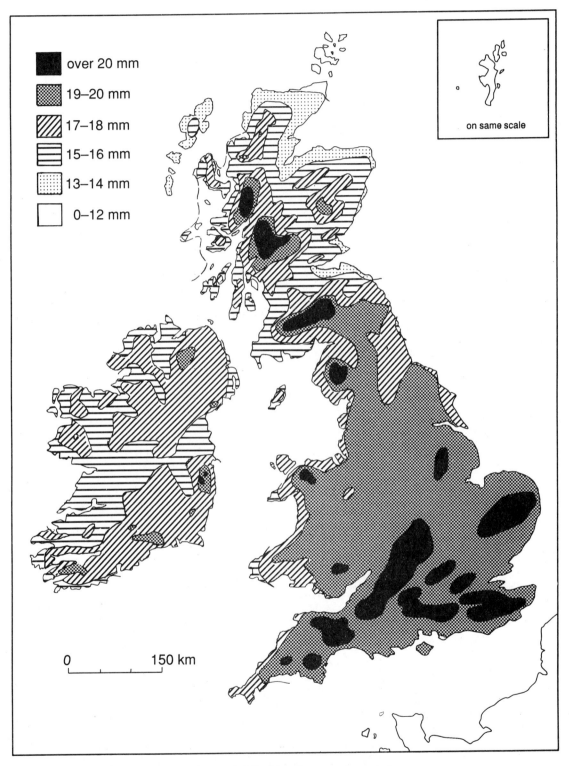

Fig. 42. 60-minute rainfall with a 5-year return period.
(Modified after Jackson 1977b, fig. 1.)

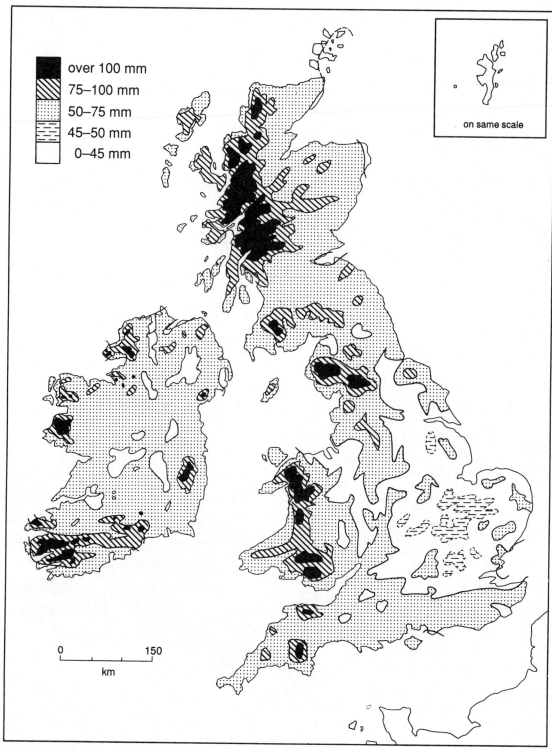

Fig. 43. 2-day rainfall with a 5-year return period.
(Modified after Jackson 1977*b*, fig. 2.)

The frequency and duration of rainfall

The frequency and duration of rainfall in the British Isles can be generalized by the expression that, on average, it rains often but not for very much of the time. The frequency of an event is the number of times that a given rainfall is equalled or exceeded. A useful simple indication of frequency is the number of rainy days that occur on average in a year (Figure 44). A rain-day is defined as a period of 24 hours commencing at 09.00 hours during which 0.2 mm or more of rain is recorded. In the far west of Ireland and both the islands and highlands of Scotland, rain may fall on or above 250 days in the year (over two days in three). In the south-east of England rain may fall on or below about 175 days in the year (just under one day in two). Inland, in central and east England, low-lying areas seem to have the lowest number of rain-days, and in the lower Thames estuary the number of rain-days is as low as 150 in the year. The available data suggest that one year in a hundred will have about 110 rain-days in the south-east of England, and about 310 in the north-west as compared with the average of 175 and 250 days. This is a measure of the frequency variation.

The amount of time that it actually rains is measured in hundreds of hours (Figure 45). For the period 1931–40, the data suggest that Great Britain actually has quite a dry climate! The area with the lowest number of hours of rain is, as one might expect, the low ground of the lower Thames Valley, East Anglia and the Fens, and parts of the West Midlands in the lee of the Welsh hills. These areas frequently have less than 500 hours per year of recorded rain. In other words, rain falls for no more than about 5 to 6% of the time. Most of England to the south and east of a line from Plymouth to the Tees has fewer than 600 hours (about 7% of the time). Almost all of Wales, all but the easternmost portions of Scotland, and much of Ireland, have mean annual totals above 700 hours (about 8% of the time). The highest values of all occur in the western Highlands of Scotland, with maxima of about 1500 hours (about 17% of the time).

By comparing the number of hours of rain in the year with annual rainfall totals, it is possible to gain an idea of the average fall of rain that takes place in an hour. This is low, and averages about 1.3 mm per hour. The figure is, of course, imaginary yet there is some truth in the idea that, for much of the time, it does not rain in Britain, it drizzles. It also suggests that much of the rain falls in discrete events, sometimes of high intensity. This suggestion is supported by the figures for the number of days with thunderstorms. The concentrations in the south-east should again be noted.

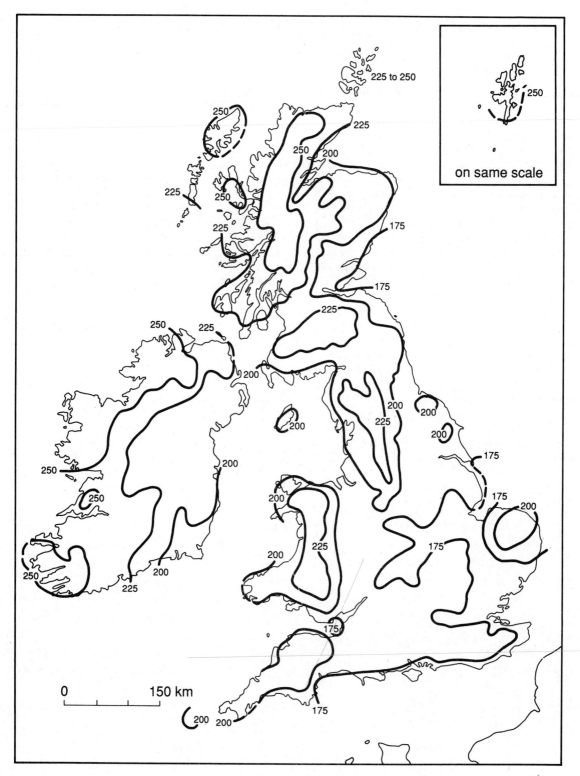

Fig. 44. Annual number of rain-days averaged over the period 1901–30.
(Based on data from the UK Meteorological Office.)

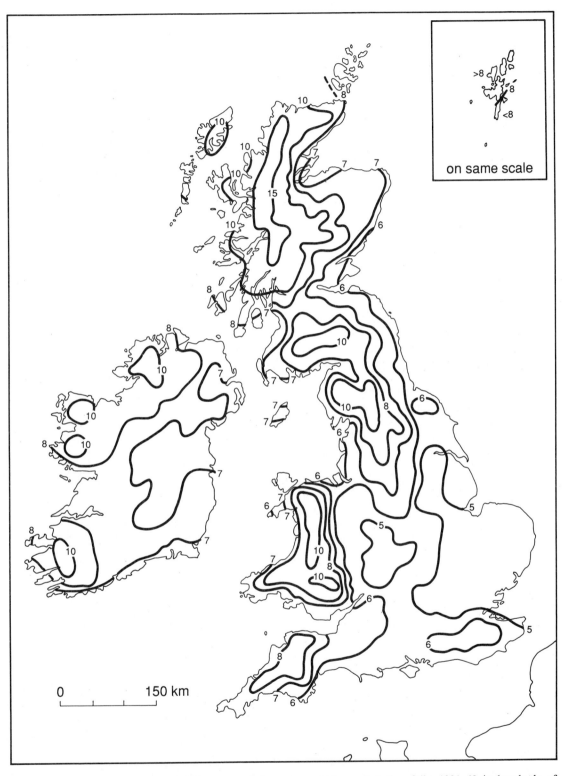

Fig. 45. Mean annual hours on which precipitation falls, 1931–40 in hundreds of hours.

(Based on data from UK Meteorological Office.)

Intense rainfall events

Although the British Isles are not subjected to such intense rainstorms as some other parts of the world, notably those parts of the tropics where intense cyclones can occur, high magnitude events do occur from time to time (Figure 46).

Daily rainfalls greater than 100 mm have been surpassed on some occasion during the period of meteorological records in almost every part of the country. While many of the heaviest falls have occurred in those parts of the country where the largest annual totals are experienced (i.e. the north and west), there have from time to time been events of a broadly similar magnitude in areas where low annual totals are the norm. This is particularly true near to the east coast (e.g. in northeast Norfolk) and in parts of the West Country. Indeed, the neck of the south-west peninsula of England (the Severn estuary area, Avon, Somerset, and Dorset) seems peculiarly prone to heavy rainfall events, perhaps because summer convectional storms are reinforced by the effects of topography. The most noteworthy example is a fall that took place at Martinstown, near Dorchester in Dorset, in July 1955, when 280 mm were recorded. This was the largest amount ever caught in a rain-gauge in one day in any station in the United Kingdom (Table 3).

The pattern of storms in Figure 46 suggests that there is an element of randomness in the distribution of events. This means that sooner or later everywhere in the UK will suffer a heavy rainfall. Such storms should also be treated as the rainfall events which cause significant landscape change because their effects often remain for a long time.

Table 3. Some heavy flood-generating rainfall events in the UK in the twentieth century

Date	Location	Rainfall
29 May 20	Louth, Lincolnshire	153 mm 3h^{-1}
17 May 36	Chilterns	42 mm
29 May 44	North-Mid-Wales	54 mm 1.5h^{-1}
15 Aug. 52	Lynmouth, Devon	>225 mm 24h^{-1}
18 July 55	Weymouth, Dorset	279 mm 6–9h^{-1}
11 June 56	Hewendon, Yorkshire	154.7 mm 1.75h^{-1}
8 June 57	Camelford, Cornwall	138 mm 2.5h^{-1}
6 Aug. 57	West Derbyshire	150 mm 5h^{-1}
8 Aug. 67	Forest of Bowland	117 mm 1.5h^{-1}
10 July 68	Mendip	101 mm 2h^{-1}
17 Jan. 74	Lomond, Strathclyde	238.4 mm 24h^{-1}
5 Aug. 78	Mid-Wales	72.7 mm 6h^{-1}
15 Aug. 77	Mid-Wales	86 mm 1.3h^{-1}
14 Aug. 75	London	171 mm 3h^{-1}
14 Aug. 75	Walshaw Dean, Yorkshire	171 mm 3h^{-1}

Source: From data in Goudie 1990; and Collier, 1990, 'Assessing and forecasting extreme weather in the United Kingdom', *Weather*, 45.

Fig. 46. Distribution of the largest daily rainfalls recorded in the UK, 1863–1960.

(*Source*: Rodda 1970, fig. 1.)

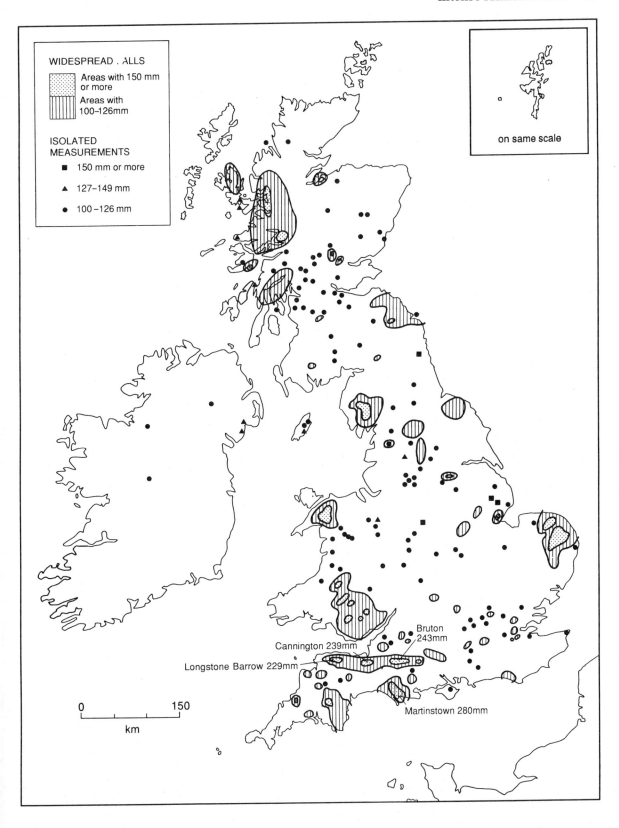

WIDESPREAD . ALLS
Areas with 150 mm or more
Areas with 100–126mm

ISOLATED MEASUREMENTS
■ 150 mm or more
▲ 127–149 mm
● 100–126 mm

on same scale

Cannington 239mm
Bruton 243mm
Longstone Barrow 229mm
Martinstown 280mm

0   150
km

Changing summer rainfall

In the 1980s and early 1990s there has been some concern that extended spells of dry summer weather have led to restrictions on water use in certain parts of the United Kingdom. Figure 47 shows summer rainfall percentages (i.e. for June, July, and August) for 1981–90 in relation to the 1951–80 average. A marked regional contrast is evident. Rainfall is about the 1951–80 average in the north-west, to the west of the topographic divide comprising the Pennines and Scottish Highlands. It is below the 1951–80 average in southern England towards the south coast. The lowest percentages (under 80%) are found around Oxford and from Dorset to Kent. These two patterns are compatible with the westerly and anticyclonic anomalies of regional airflow found over Scotland and southern England respectively since the 1970s. This pattern seems to have continued until the 1990s and has been accompanied by four of the hottest and driest years on record. It is too early to say whether this is part of a wider pattern and far too easy to regard it as part of a global-warming pattern. It does, however, have severe implications for water-supply planning in the United Kingdom. Most of the demand is in the south, south-east, and midlands whilst increasingly the supply of excess water is in the north and west. With privatization of the water companies costs might be expected to rise in the demand regions and should fall in the water-rich areas as water is purchased and transferred across the country. Will this be reflected in future share prices and dividends! A further concern is that the summer supply is crucial in the southern resort towns. The influx of visitors means that demand can triple in the summer months so that a summer excess-supply capacity must be built in.

All this means difficult planning decisions and a need for a real understanding of the physical geography as well as the socio-economics of the country. For example, by 2021 it is anticipated that, if all the projected water-supply schemes are completed, water companies like Northumbria will have an excess of 800 million litres a day, the North-west over 550 million litres a day, and Wales 300. Thames and Southern will barely have enough, despite the recharge of the chalk aquifer, and Anglia will be in deficit by over 100 million litres a day. The current dry spells mean that the normal 15% safety margin is reduced to 3%. Hosepipe bans save only 3%, so the situation is critical.

A final consideration is that if, as the companies believe, the 'drought' is temporary they may find themselves investing in new schemes which become surplus to requirements when the 'dry' trend changes. Shareholders may not understand or approve. It is almost certain, therefore, that an alternative strategy will be adopted, namely pricing and domestic metering of water used. There is no better way of reducing demand than asking individuals to pay for a product that they have previously had for free.

Fig. 47. Summer rainfall (June, July, and August) in the UK in 1981–90 as a percentage of that for the period 1951–80.

(*Source*: Mayes 1991, fig. 2.)

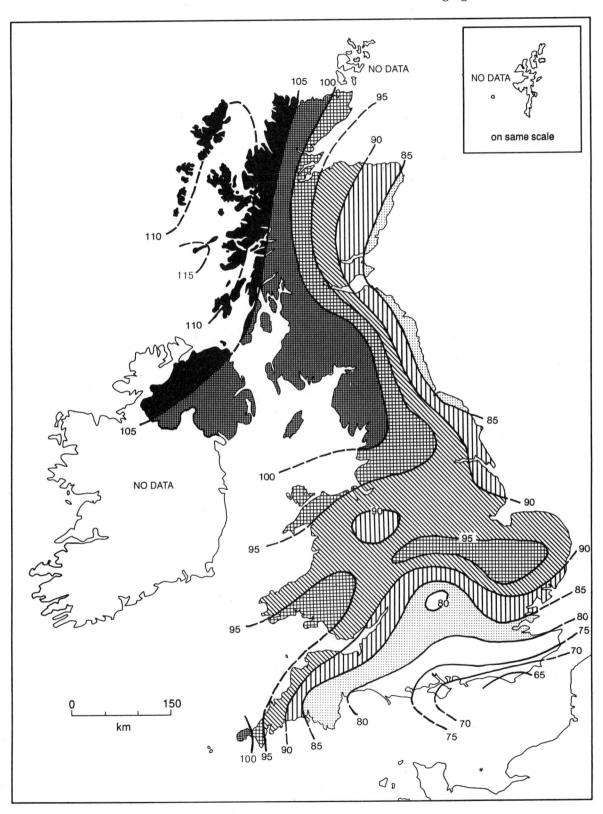

NO DATA

105 100

95

90

85

NO DATA

on same scale

110

115

110

105

NO DATA

100

90

95

90

95

90

95

90

85

80

80

75

70

65

0   150

km

95

80

70

75

100 95 90 85

Snowfall

Heavy snowfalls are not common in the British Isles, and E. G. Bilham, in his *The climate of the British Isles*[1] (1938) remarked, 'the younger generation of today may be excused for wondering why snow should be associated so closely in British folklore and tradition with winter, and particularly with Christmas. There can be no doubt that we have here a survival of recollections of days when our winter climate was definitely more severe; when it was natural for a King Wenceslas or an outcast heroine to face snow instead of rain or fog.' The generation of 1994 may not see a white Christmas and certainly don't know who Bing Crosby was!

Figure 48 shows the number of hours and Figure 49 the number of days with snow lying in an average (median) winter if the effect of altitude is removed. There is a fairly simple pattern with a gradient from south-west to north-east. In upland areas like the northern Pennines or the Highlands of Scotland there may be more than 60 days in the year with snow lying on the ground. On the mild south-west coasts lying snow is a relatively rare phenomenon, for the frequency at Valentia (south-west Ireland) and Falmouth (Cornwall) is no more than 1 day per 2 years. Figure 50 shows the depth or level snow reached or exceeded in 10 winters per century. In the more easterly highland regions of England and Scotland values of over 50 cm are recorded. In the British Isles there is a clear relationship between snow cover and altitude. The rate of increase of snow cover with elevation ranges between 5 and 15 days per 100 m rise. The lower rate is applicable to the moorlands of Devonshire and the higher rate to the central Highlands of Scotland. Figure 49, showing the mean annual number of days with snow lying at 9.0 ma. (1941–70), indicates a similar pattern. Snow lies deepest, longest, and for most days in Scotland, Wales, and the northern uplands. The coast of the south-west, south Wales, and the Bristol Channel is mild in all senses.

[1] Macmillan: London.

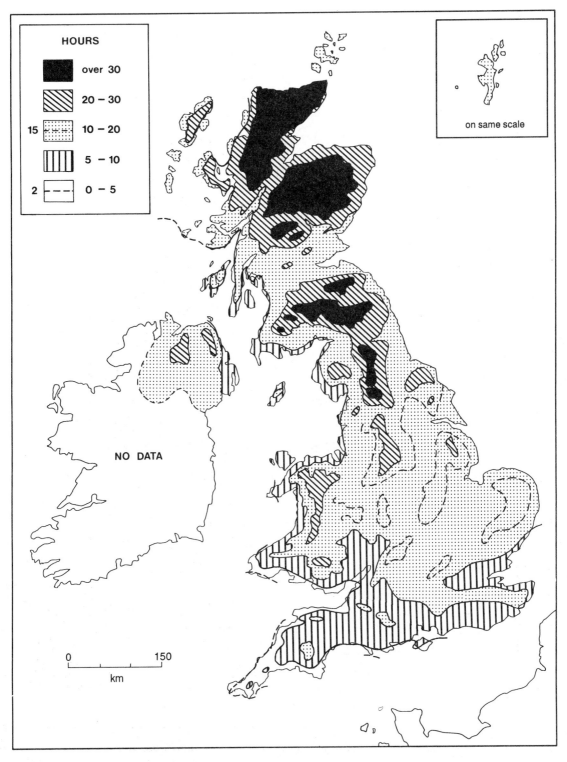

Fig. 48. Estimated mean annual number of hours of moderate or heavy snow. (Modified after Jackson 1977*a*, fig. 4.)

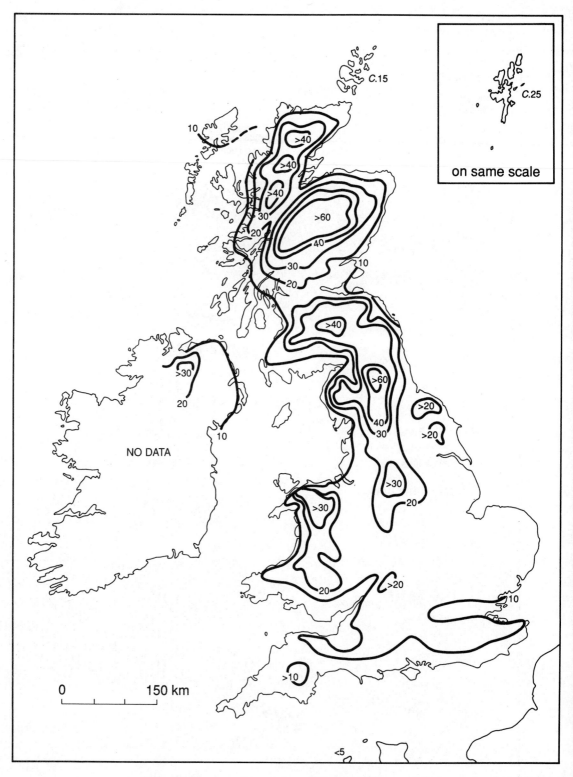

Fig. 49. Mean number of days with snow lying at 09.00 hrs each winter, for the period 1941–70.

(*Source*: Perry 1981, fig. 3.5.)

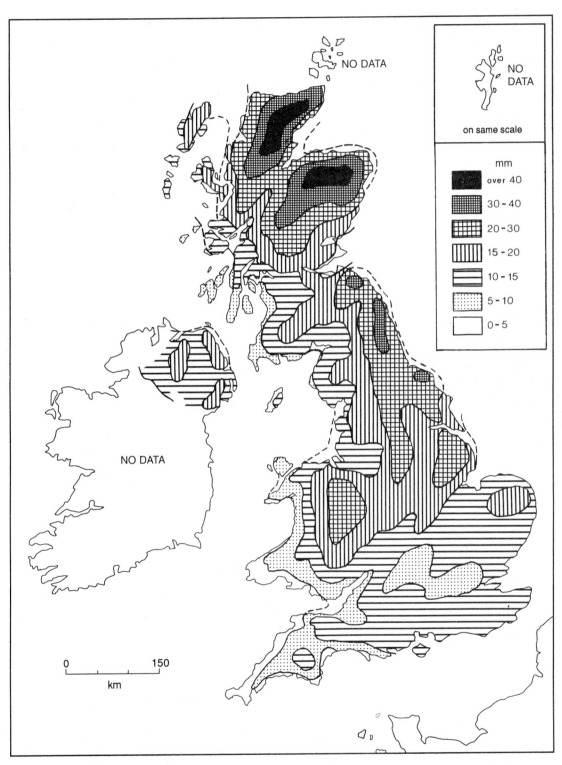

Fig. 50. The 5-year water equivalent of lying snow (mm) reduced to sea-level.

Relative humidity

Water is present in the air in varying amounts; the measure of it is referred to as humidity. At any specified temperature the quantity of moisture that can be held by the air has a definite limit, with warm air being able to hold more than cold. The proportion of water vapour actually present at a given temperature, relative to the maximum quantity that could be present, is called the relative humidity.

It is also possible to express humidity in terms of the actual quantities of moisture in the air. There are a variety of indices that are used. One of these is absolute humidity, the mass or weight of water vapour per unit volume of moist air. Another is vapour pressure, which is the partial gaseous pressure exerted by the water vapour (normally expressed in millibars).

Figures 51, 52, and 53 show the distribution of relative humidity at 15.00 hours for January, July, and the whole year respectively, based on readings for the period 1961–70. Over the year, and particularly in July, the increasing diurnal range of temperature with distance from the sea exerts a dominating influence on mid-afternoon values and results in an inland decrease in values. The contrast is less marked in Ireland.

Moreover, as the January values show, during the winter months the contrast between coastal and inland sites becomes less marked, and there are only small variations in relative humidity from place to place. Other factors, such as air-mass type and prevailing wind assume greater importance.

Figure 54 shows the average number of hours with a relative humidity at 90% or more. Areas on the eastern side of high ground have noticeably shorter periods of time with high values of relative humidity. Of particular interest here are the low values around the Liverpool district, which is in the lee of both the Welsh hills and the Pennine Range (according to wind direction), and on the shores of the Moray Firth in north-east Scotland. These favoured areas have respectively about one-quarter and one-third of all hours with relative humidity of 90% or above, whereas on higher land the duration of such humid conditions is much greater, and may well be over 50% in some upland regions. It is also interesting that some areas of the south-west maintain high humidities despite their position in the lee of Dartmoor. This may be related to the pattern of intense summer storms and to the frequency of sea mist.

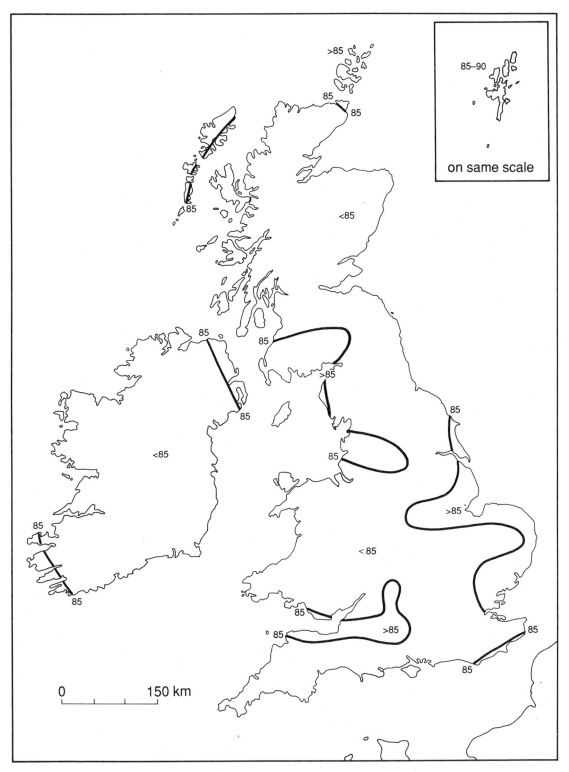

Fig. 51. Mean January values at 15.00 hrs, for the period 1961–70.
(*Source*: UK Meteorological Office Data.)

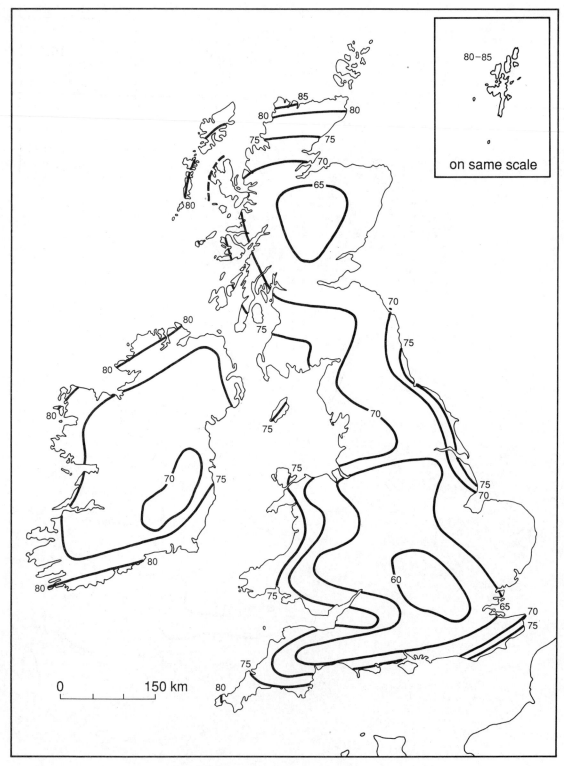

Fig. 52. Mean July values at 15.00 hrs, for the period 1961–70.
(*Source*: UK Meteorological Office Data.)

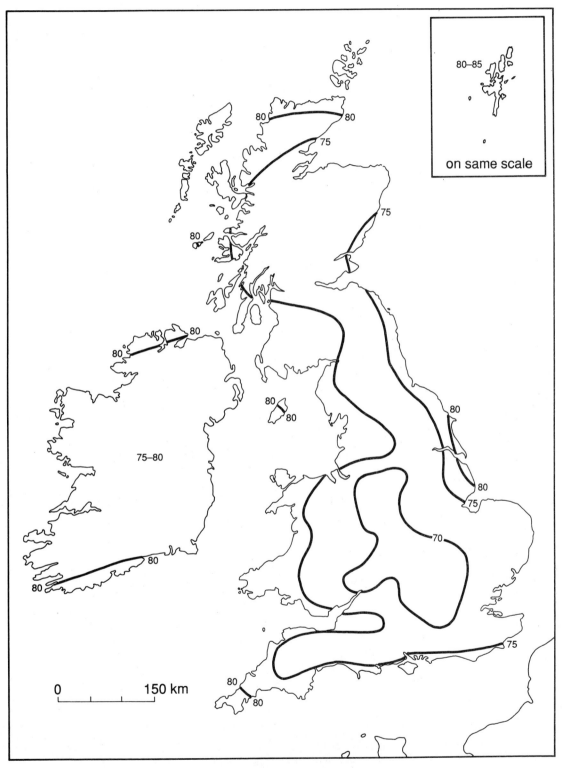

Fig. 53. Mean annual values at 15.00 hrs, for the period 1961–70.

(*Source*: UK Meteorological Office Data.)

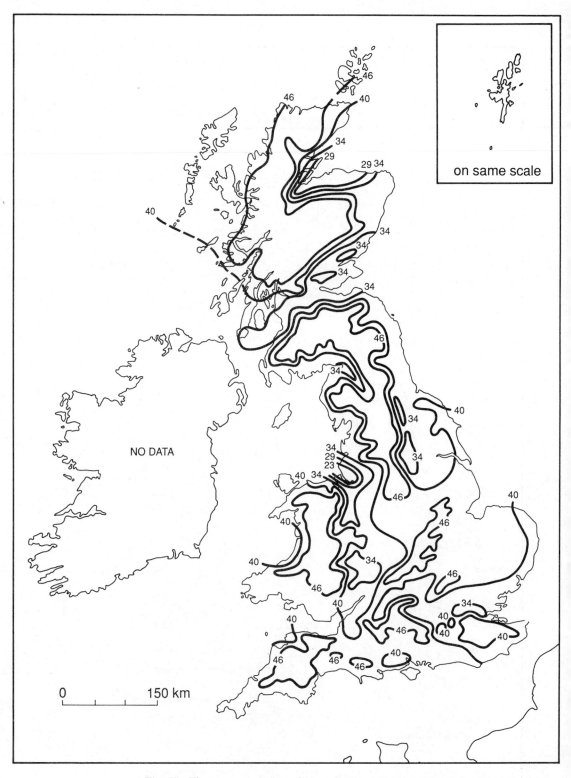

Fig. 54. The average number of hours in Great Britain with relative humidity 90% or more (shown as a percentage of all hours).

(*Source*: UK Meteorological Office Data.)

Evaporation

Moisture can be lost from drainage basins because of the process of evaporation. This involves the loss of water as vapour from the surface of water bodies, from soil, from plant leaves (transpiration) and from the wetted surfaces of vegetation (interception loss). It is necessary to draw a distinction between two measures of evaporation. Potential evaporation or evapotranspiration is the theoretical *maximum* water loss which would occur in given climatological circumstances from a moist surface. It depends on such crucial factors as temperature, solar radiation, and wind. Real or actual evapotranspiration is the water loss which takes place from vegetation and soil surfaces which dry out from time to time, and from which, as a result, little or no actual evapotranspiration can actually take place at certain periods. The extent to which there is a difference between potential and actual evapotranspiration in different parts of the year determines the amount and the duration of times of water surplus or water deficit, which in turn have great implications for water management, agriculture, and so forth.

Figure 55 shows mean annual potential evapotranspiration for the British Isles, based on data from the 1960s. It shows a broad latitudinal decrease of potential evapotranspiration from south to north and a marked decrease away from the coast. The highest values, around 550–600 mm, occur around the coast from west Wales to Suffolk, and in the vicinity of the Dee and Mersey estuaries in north-west England. It is striking that this broadly reflects the microclimates noted on the relative humidity maps. The lowest values, below 400 mm per annum, occur over the central Highland areas of Scotland.

Figure 56 is an isopleth map of actual evapotranspiration (AE), showing values in the south and east which are higher than those in the north and west. In much of England the values are between about 400 and 500 mm, and this in turn means that about half of mean annual rainfall is 'lost' because of this process. Elsewhere in the British Isles the proportion is greatly diminished. In the high rainfall areas of the uplands of Scotland, where actual evapotranspiration losses may be 300 mm or less per year, the percentage of precipitation 'lost' may fall to less than 20%. In south-east England, particularly to the east of a line linking Dorset and Yorkshire, there may be a considerable deficit of soil moisture in the summer months, creating water-supply difficulties, and making irrigation a desirable basis for the successful cultivation of certain crops. These data should be compared with the maps of rainfall variability and summer rain for they emphasize what controls the water supply, and the need for the planning system to understand fully the physical geography of the country.

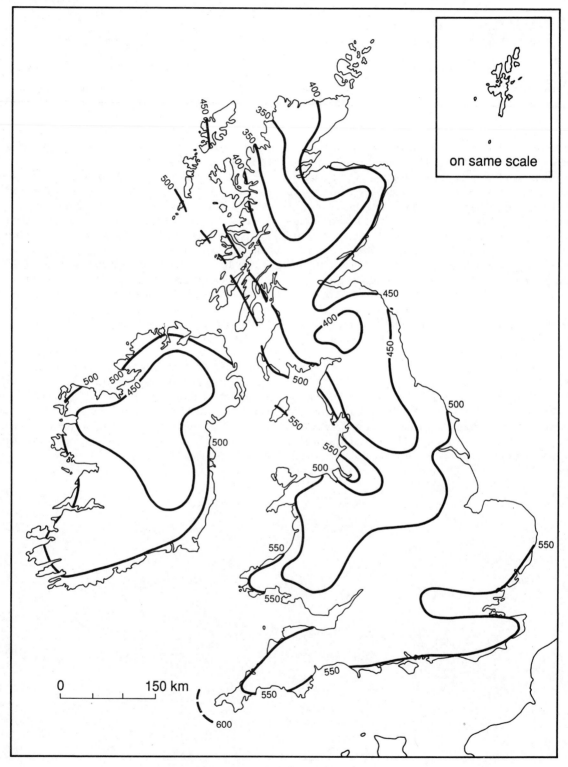

Fig. 55. Mean annual potential evapotranspiration (PE) in mm.

(*Source*: Chandler and Gregory 1976, fig. 7.1.)

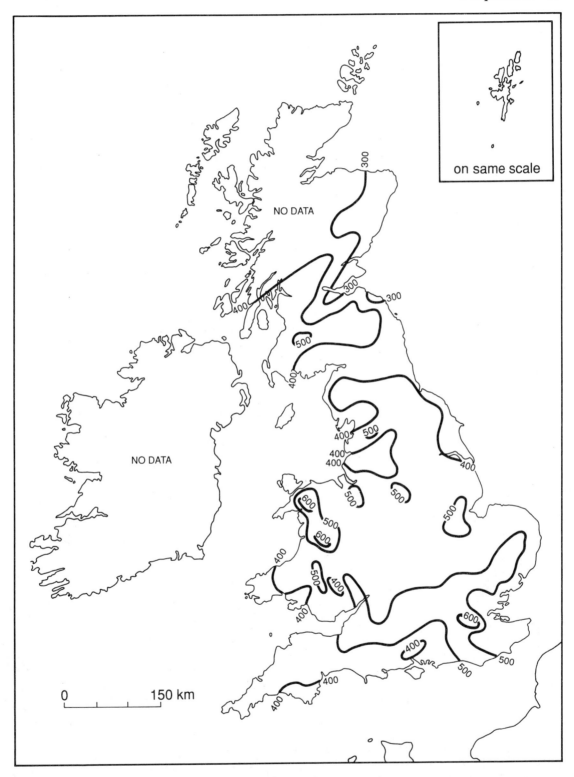

Fig. 56. Mean annual evapotranspiration (AE) in mm.

(*Source*: Chandler and Gregory 1976, fig. 7.2.)

Cloud cover

The British Isles have sometimes been called 'the elusive archipelago', because from space they are often obscured by cloud. Maritime influences and a succession of mid-latitude depressions moving in from the Atlantic Ocean mean that the area is often very cloudy.

Cloud observations are reported in 'oktas', the eighths of the sky covered by cloud. An absence of cloud is denoted by 0, the presence and amount of cloud by values between 1 and 8, and an obscured sky (resulting from darkness, fog, or smoke) by a value of 9.

Figure 57 shows that a substantial cloud cover (7–9 oktas, occurs most frequently in the north and west of the British Isles, especially in upland areas like the Lake District, and the Southern Uplands and Highlands of Scotland. Cloudiness is generally enhanced over high ground, and especially on the exposed windward sides

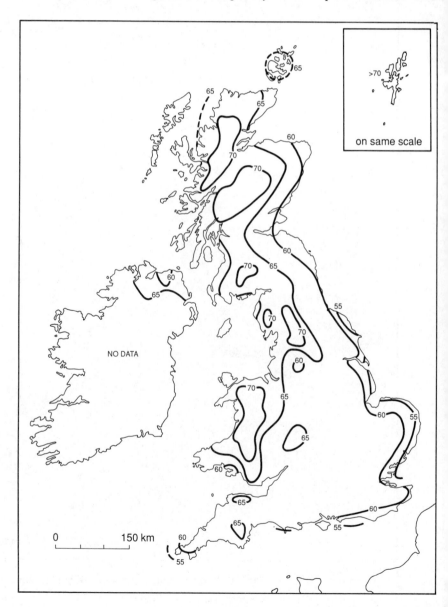

Fig. 57. Percentage frequency of days with 7–9 oktas cloud cover at 09.00 hrs, for the period 1957–70.

(*Source*: data from UK Meteorological Office.)

of mountain ranges. It is less over low-lying ground and, rather strikingly, along coasts, particularly if they are sheltered from the west. Indeed, as in many of the maps the parallelism between the pattern of isopleths of overcast conditions depicted in Figure 57 and the configuration of the coastline is striking and revealing. It is for this reason that the Outer Hebrides have much the same number of cloudy days as do some of the more inland areas of southern England. This general pattern probably serves to emphasize the great significance of cloud enhancement that takes place over the land as a result of orographic uplift and stronger convection.

The pattern in Figure 58 is for the most part the inverse of that in Figure 57, for it shows the percentage of days when cloud cover is absent or low (0–2 oktas). The pattern, however, is by no means an exact mirror image. Most especially, it is evident that the south-east of England has a large number of days with little or no cloud cover in comparison with the north and west. The small area in central-western England with a relatively high incidence of clear skies may perhaps be explained by a rain-shadow effect caused by the subsidence of relatively dry air in the lee of the Welsh mountains. This drier area is also shown on the maps of annual rainfall (Figure 40), annual hours on which precipitation falls (Figure 45), and the distribution of large rainfalls (Figure 46).

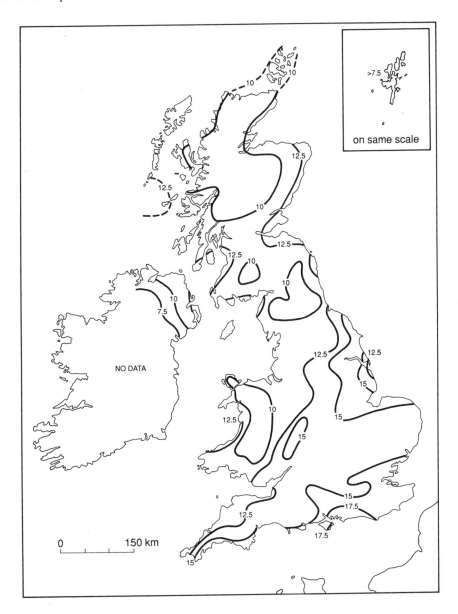

Fig. 58. Percentage frequency of days with 0–2 oktas cloud cover at 09.00 hrs for the period 1957–70.

(*Source*: data from UK Meteorological Office.)

Droughts

Although the British Isles tend to enjoy a relatively reliable rainfall regime, there are occasions when droughts occur. These droughts are notable because they cause water shortages, increase the danger of fire, may reduce crop yields, and cause soils to crack, thereby damaging the foundations of buildings built on clayey substrata. Notable droughts of recent years have been those of 1975–6, and those of 1988–92.

Two sets of circumstances tend to give rise to drought conditions: an exceptional persistence of blocking anticyclonic conditions, which cause the diversion of rain-bearing depressions away from the British Isles, or the reversal, for long periods, of the normal pressure gradient so that the flow of air comes from the east rather than the west.

The definition of drought used by meteorologists in Britain is that an *absolute drought* is a period of at least 15 consecutive days on which no more than 0.25 mm of rain falls, while a *partial drought* is a period of at least 29 consecutive days, the mean daily rainfall of which does not exceed 0.25 mm.

Figure 59 demonstrates that for the period 1906–40 the average number of absolute droughts was more than one per year in southern England and less than one in every five years in northern Scotland. This suggests that dryness can be a significant hazard in the UK. This is probably true with respect to plant growth and the need for irrigation. It is perhaps less true in a

geomorphological sense for which the 15-day definition is less applicable.

The drought of 1975–6 was an event of extreme severity and climatologists calculated that such an event has a return interval of only once in perhaps 500 years (Figure 60). Rainfall was less than 60% of average over a wide area of England between Devon and the Humber, and it was only in a very limited portion of north-west Scotland that precipitation totals were normal. Figure 60 should be compared with the map of the coefficient of variability (Figure 41). This suggests that the bad drought periods are part of a long-term characteristic of the British climate, rare but not unexpected!

Another major drought occurred in 1989. The rainfall map for the period November 1988 to November 1989 shows the rainfall quantity over that period as a percentage of the corresponding 1941–70 average (Figure 61). A rather similar pattern to that experienced in 1976, is revealed with normal or above normal values in the north-west of Scotland, but with values less than 60% of average in the east of the country (including, this time, the east of Scotland). This dry interval also seems to be part of a generally dry period extending from 1988–92. It is interesting to speculate whether this is part of a regular cycle of wet and dry periods or whether there is a new trend associated with global environmental change.

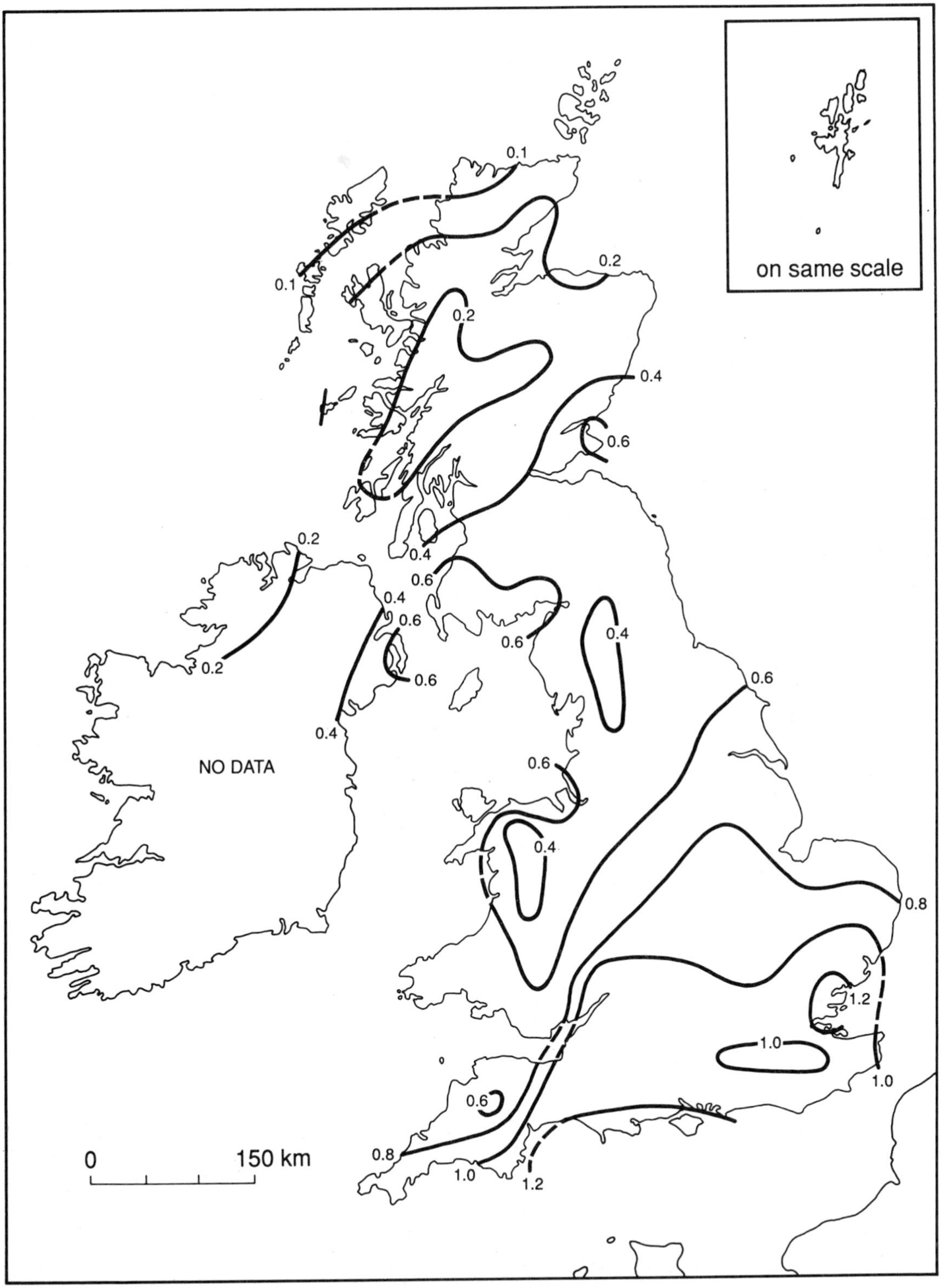

Fig. 59. The regional distribution of mean annual absolute drought.

(After Glasspoole and Rowsell 1947, in Perry 1981, fig. 5.1.)

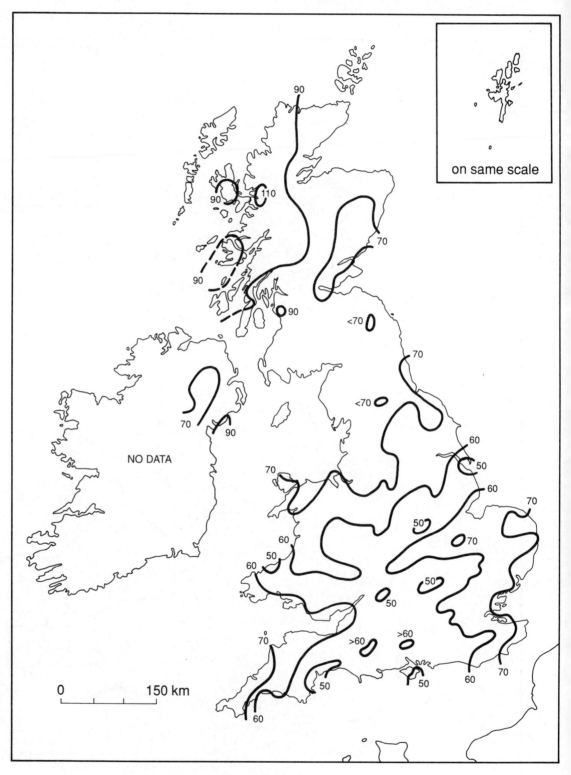

Fig. 60. The spatial distribution of UK rainfall for the period May 1975–Aug. 1976 as a percentage of the 1916–50 mean.

(*Source*: Perry 1981, fig. 5.2.)

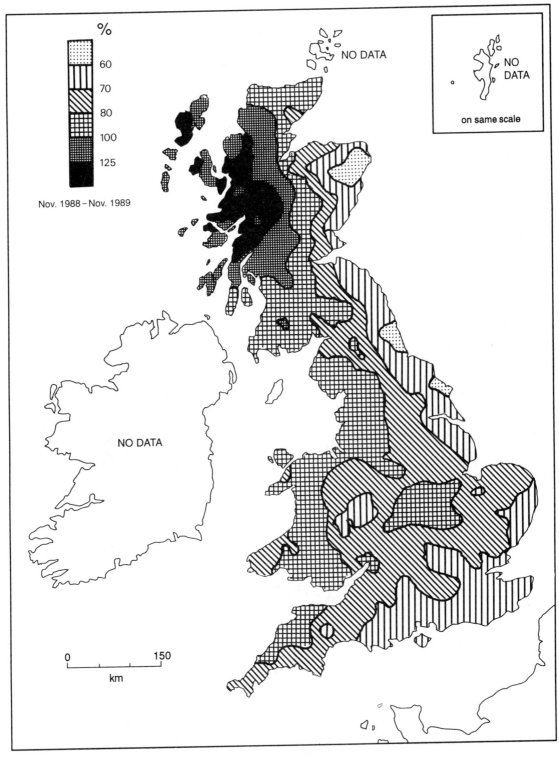

Fig. 61. Great Britain rainfall for the period Nov. 1988 to Nov. 1989 as a percentage of the corresponding 1941–70 average.

(*Source*: Marsh and Monkhouse 1990, fig. 3.)

Wind direction and frequency

The important characteristics of wind are direction, force, and frequency of velocity. Wind direction is recorded on maps in the form of a series of twelve-point wind-roses based on 30 degree direction ranges. Force is shown by the thickness of the arrow and is given in a series of groups: 1 to 3 (light winds); 4 (moderate winds); 5 and 6 (fresh to strong winds); and 7 and above (near gales and gales). The percentage of calms is given in the centre of the rose, and the length of each part of an arrow, which flies with the wind, is proportional to the frequency of speeds in the range which it represents. Speeds relate to a standard height of 10 m above the ground.

In general the most common or prevailing wind directions are from the south and west. The least frequent winds are those having easterly components. The frequencies of winds of force 5 and over on the Beaufort scale are much greater in the north and west than they are in the south and east. Calms and light winds are more common in the east than in the west, while winds of force 7 or over are relatively infrequent in England (Figure 62).

Comparable diagrams can be drawn to show the differences between winter (Figure 63) and summer (Figure 64). The January wind-roses also suggest the dominance of winds from the south and west, though there are fewer winds from between north and east and more

winds from between south and west than there are over the entire year. Winds of force 5 and over are, however, more frequent than they are over the year as a whole. The July wind-roses show a comparable directional pattern. The frequencies of light winds are relatively greater and those of strong winds relatively less than they are in January or over the year as a whole.

Figure 65 shows the annual percentage frequency of wind speeds from all directions. The radius of the central circle around each station gives the frequency of winds under 4 knots and the widths of the three successive outer zones give the percentages falling within the ranges 4–10, 11–21, and 22 or more knots. The highest annual frequency of strong winds and gales are recorded at Tiree, Lerwick, Ronaldsway, Belmullet, Scilly, and Dungeness. Stations having the fewest strong winds and gales are Mildenhall, Hurn, and Heathrow (all in the south and east of lowland England).

A nice contrast is provided by the map of the frequency of calm periods (Figure 65). A calm is a period in which winds do not exceed 6 knots. The relief effect is again obvious with more than 40% of annual calms occurring in lowland Britain and perhaps surprisingly, to a southerner, in eastern Scotland! The windiness of the east and west coasts, but the relative calm of the south coast between Dorset and Sussex, is clearly portrayed.

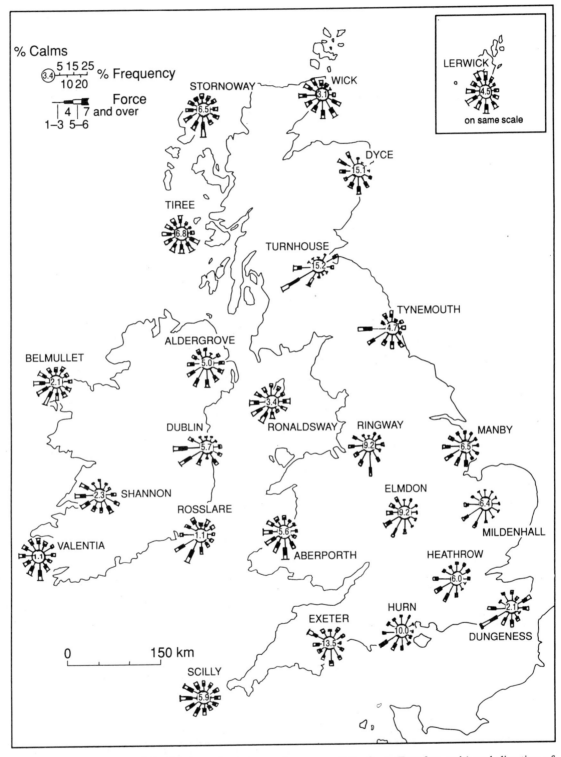

Fig. 62. Annual percentage frequency of the force (Beaufort scale) and direction of the wind.

(*Source*: Chandler and Gregory 1976, fig. 3.1.)

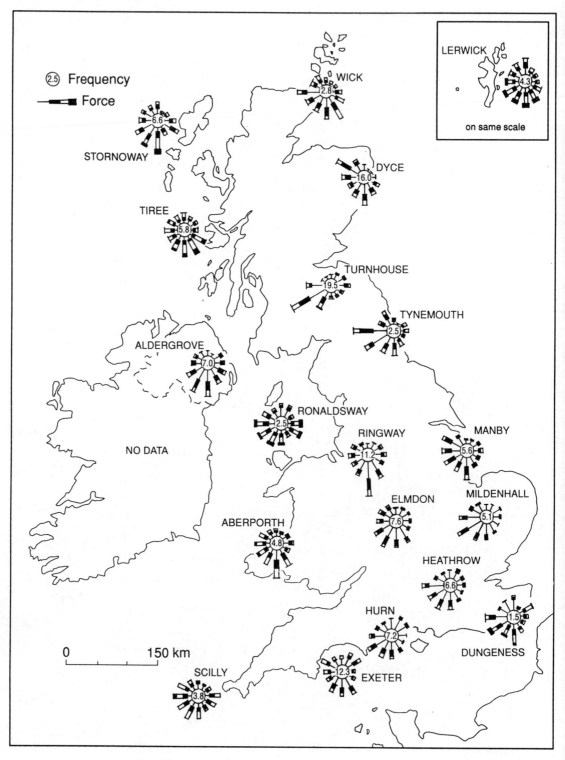

Fig. 63. January percentage frequency of the force (Beaufort scale) and direction of the wind.

(*Source*: Chandler and Gregory 1976, fig. 3.2.)

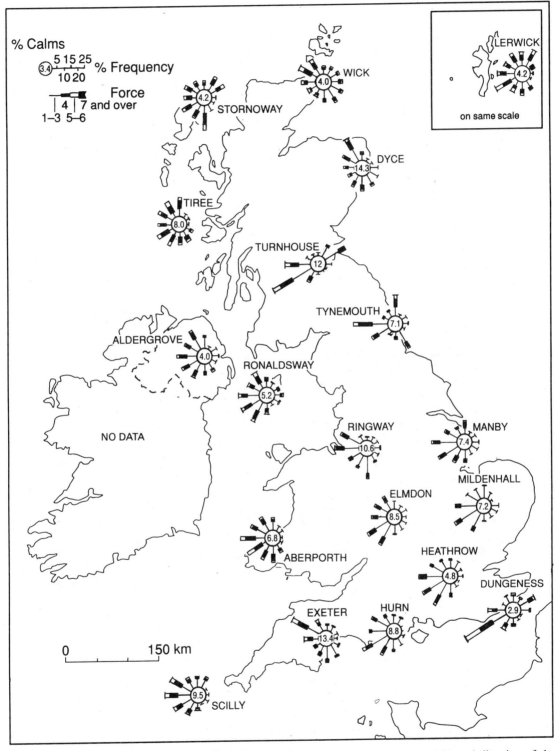

Fig. 64. July percentage frequency of the force (Beaufort scale) and direction of the wind.

(*Source*: Chandler and Gregory 1976, fig. 3.3.)

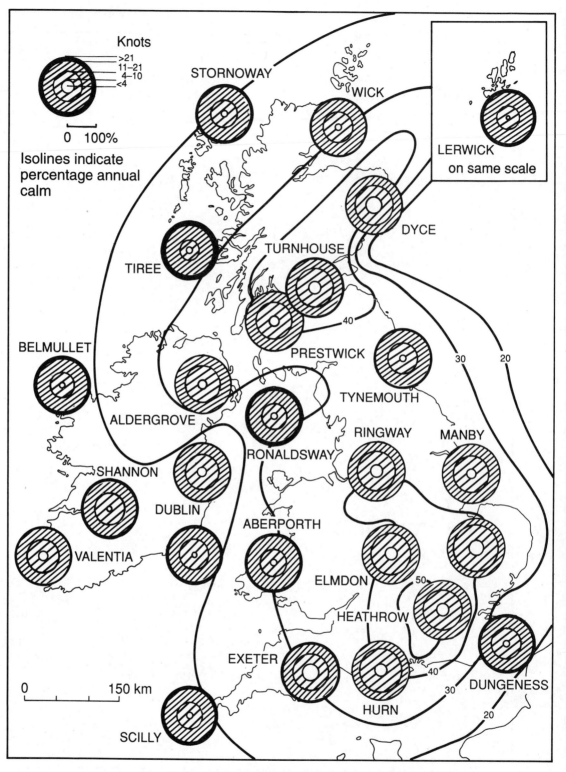

Fig. 65. Annual percentage frequency of mean wind speed from all directions. The radius of the outer circle represents 100%.

(*Source*: Chandler and Gregory 1976, fig. 3.15); with superimposed the percentage of annual calm (including frequencies of winds up to 6 knots). (*Source*: Allen and Bird 1977, fig. 2.)

Strong winds

Figure 66 shows the approximate average number of days when gales occur. If the mean wind speed exceeds 17.2 m per second for at least 10 minutes, then this is regarded by the Meteorological Office as a 'day with gale'. Proximity to the coast seems to be a major control of the distribution pattern. Thus, whereas in much of inland Britain days with gales occur on less than 5 days in the year, in western, coastal locations (e.g. the Orkneys and Shetlands, the Hebrides, Pembroke, and the south-western tip of Cornwall) there may be 30 or more days in the year with gales. The map applies to open country areas with an altitude of less than 200 m. True values will be more for exposed mountain sites and less than indicated in urban areas.

It is also important to know the wind speeds that may be attained in gusts. Figure 67 shows the hourly wind speed with a recurrence interval of 50 years over open country, while Figure 68 shows the maximum gust speed that may be anticipated once in 50 years. The range of once-in-50-years hourly mean wind speeds varies from only 32 knots at Kew (west London) to 72 knots at Lerwick (Shetland Islands). Maximum gust speeds, with a recurrence interval of 50 years, range from less than 75 knots in the interior of south-east England to over 105 in the far north and western isles of Scotland.

Both patterns are reflected in the map of available wind energy (Figure 69). Maximum energy occurs around the coast with a very steep gradient on the Scottish east coast near Aberdeen.

Occasionally very strong gusts can occur. One of the most ferocious storms to afflict southern Britain was that which took place on 15–16 October 1987 (Figure 71). At least 18 people lost their lives as a direct result of the gale, many buildings were damaged, and millions of trees were blown down. The winds were associated with the movement of a depression from south-west to north-east. Wind speeds exceeded 80 knots over a large area, and a maximum gust of 106 knots was recorded at Gorleston.

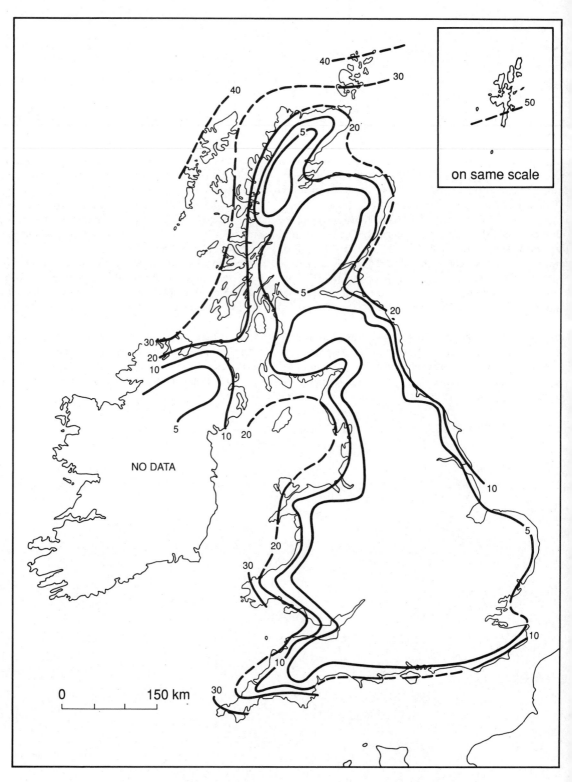

Fig. 66. The average number of days with gale (i.e. days during which the mean wind speed reaches or exceeds 17.2 metres per second for at least 10 minutes).

(*Source*: modified after Lacy 1977, fig. 40.)

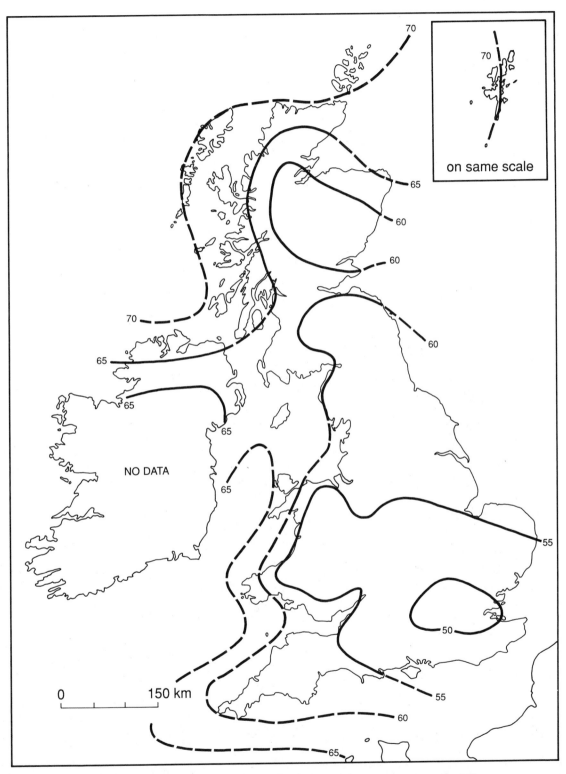

Fig. 67. Hourly mean wind speed (in knots) with a recurrence of 50 years.
(After Hardman 1953, in Perry 1981, fig. 2.5.)

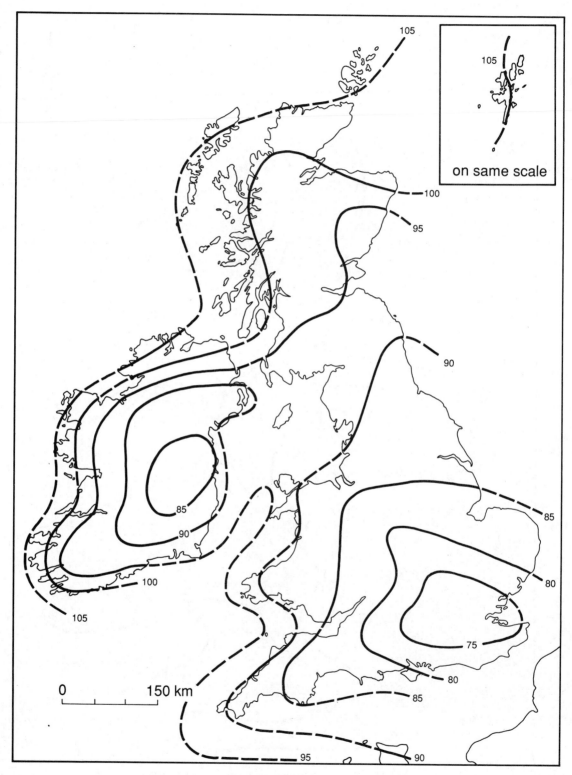

Fig. 68. Gust speed with a recurrence of 50 years.
(After Hardman 1953, in Perry 1981, fig. 2.5.)

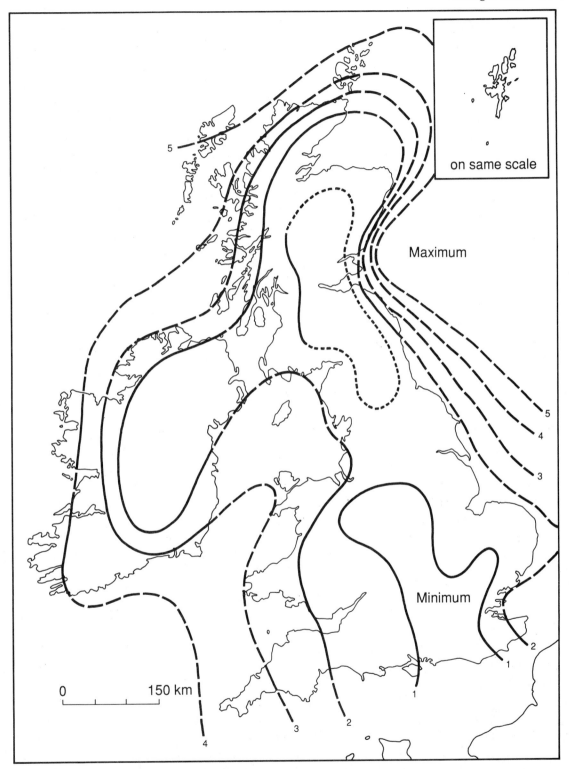

Fig. 69. Available wind energy (MWh m²) (hilltop sites not included).
(*Source*: modified after Allen and Bird 1977, fig. 1.)

Tornadoes

A tornado is a relatively narrow vortex (a few metres to hundreds of metres in diameter) of rotating winds which spiral inwards and upwards to the base of a convective cloud. They are typically short-lived (often lasting only a few minutes) and of small areal extent (track lengths being typically a kilometre or less, and widths 50 to 100 m). However, they cause great damage as a result of their swirling winds, suction, and shearing effects. The larger tornadoes are related to the passage of well-defined fronts or troughs associated with rapidly deep-ening or very deep depressions. Although tornadoes can occur in any season, the greatest frequency is in autumn and early winter, when the presence of relatively warm seas surrounding the British Isles enhances instability as cold air sweeps south-eastward or eastward to Britain. The lowest frequencies, by contrast, occur in the period from February to May, when the British seas are normally at their coldest.

Over the period 1960–82 (Figure 70), 97% of the observed tornadoes for the UK occurred in England

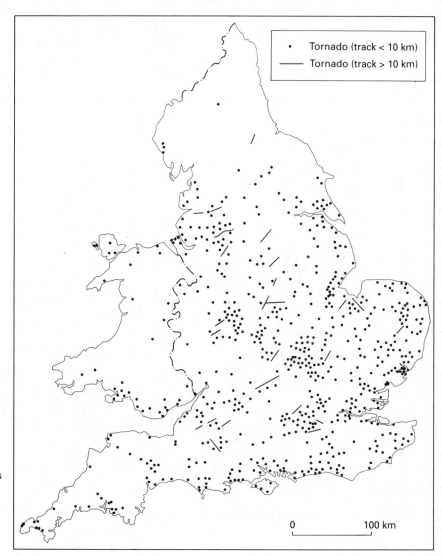

Fig. 70. Geographical distribution of tornadoes, for the period 1960–82. An arbitrary path track of 10 km is employed to distinguish between localized and long-track tornadoes.

(*Source*: Elsom and Meaden 1984, fig. 3.)

and Wales. They tend to concentrate in the lowland parts of Britain and show a very similar pattern to the convective storms. This distribution also contrasts with the map of calms (Figure 65) to stop Sussex from becoming complacent!

Some authors confuse the true tornado with very strong gales or storms. The word 'hurricane' may also be used. The storm tracks shown in Figure 71 represent extreme gusts and the main paths of a storm as indicated by records of high winds. There are often conflicting perceptions of these events. The tracks for 16 October 1987, shown here, should be compared with the centres of highest gusts which clearly shows maximum speeds centred on the Brittany, Cotentin, Normandy, Kent, Suffolk axis!

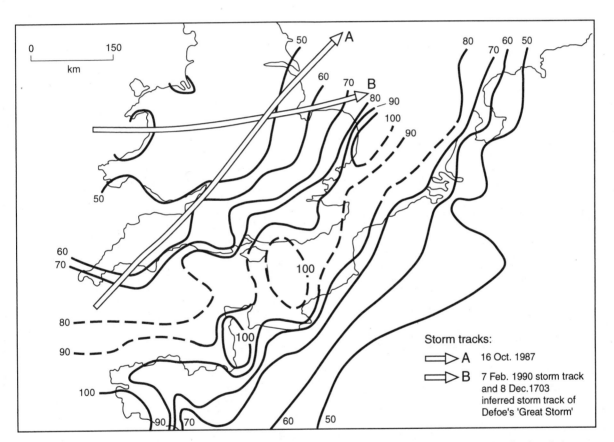

Fig. 71. The highest reported gusts (in knots) over southern England and the near-continent during the Great Storm of 16 October 1987).

(*Source*: Burt and Mansfield 1988).

It also indicates the tracks of some of the violent winter storms that have swept across the country.

(*Source*: Lewis 1991: 41.)

Solar radiation and sunshine

Solar radiation is the electromagnetic radiation emitted by the sun and intercepted by the earth, and is expressed as megajoules per square metre (MJ m^{-2}). It varies according to the elevation of the sun (which in turn is related to the season and latitude) and to the cloud and aerosol content of the sky.

Figure 72 shows the average daily total of global solar radiation for the year. As might be expected the amounts decrease from south to north (i.e. with latitude), but for any given latitude the amounts tend to be higher in the west. Areas of lower average radiation occur in the London region, perhaps because of pollution; the Midlands, Yorkshire, and the Central Lowlands of Scotland, due to cloud cover.

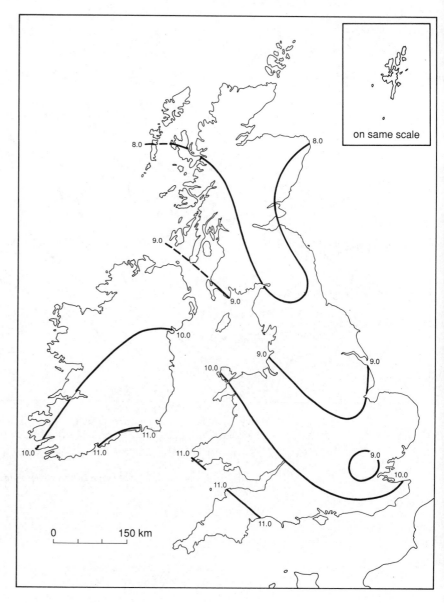

Fig. 72. Average total of daily solar radiation (MJ m^{-2}) for the year.

(*Source:* Chandler and Gregory 1976, fig. 4.8.)

Figure 73 shows the pattern of sunshine, expressed as the average daily bright sunshine in hours. In parts of the Highlands of Scotland the value is less than 3. In Ireland only the easternmost part have values over 4. By contrast, the sunniest parts of the British Isles are along the south coast of England, where, locally there are values greater than 5 hours per day. The sunniest place is probably Ventnor-Shanklin, Isle of Wight.

As one would anticipate, the amount of bright sun-shine shows a strong seasonal pattern (Figure 74). In June (Figure 74b) even the Highlands of Scotland may average not far short of 5 hours per day, while on the south coast of England the average figure may exceed 8.5 hours. By contrast, in December (Figure 74d) there are extensive areas of the Scottish Highlands which receive less than 0.5 hours of sunshine per day, and even on the south coast of England, the average only just reaches about 2 hours.

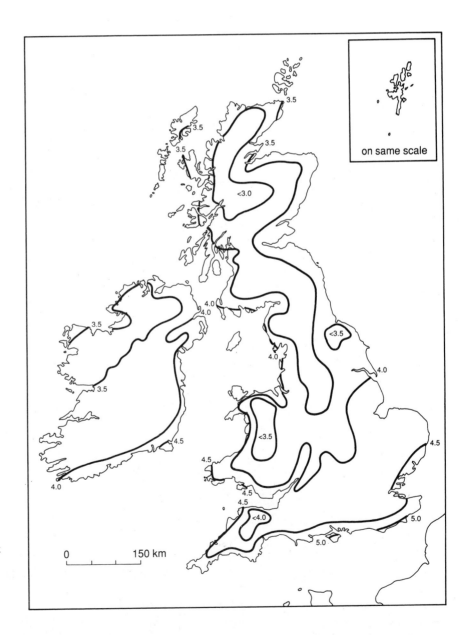

Fig. 73. Average daily bright sunshine in hours for the period 1941–70 for the year.

(*Source*: Chandler and Gregory 1976, fig. 4.13.)

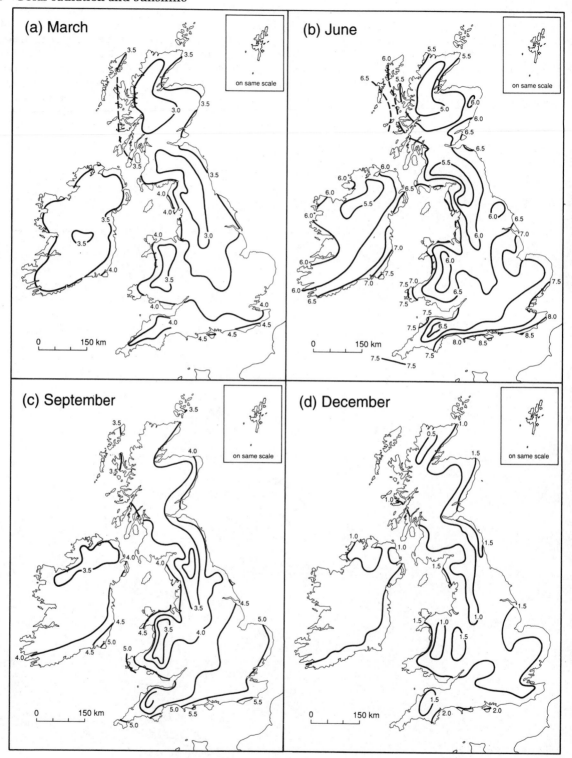

Fig. 74. Average daily bright sunshine in hours for the period 1941–70. (*a*) March (*b*) June (*c*) September (*d*) December.

(*Source*: Chandler and Gregory 1976, fig. 4.12.)

Visibility

Visibility data (Figure 75) show important geographical patterns due to meteorological causes, natural variations in topography, and to urban and industrial pollution. In the low-lying areas, with an altitude of less than about 150 m, the foggiest zones are the lowlands of Scotland between the Clyde and Forth estuaries, and a zone that runs up from south-east England across the Midlands to south Lancashire, and West Yorkshire to the north-east. The least foggy areas are the extreme northern areas of Scotland and western Ireland. These least foggy areas have high wind velocities (Figure 63) and are far removed from sources of atmospheric pollution. However, height is also a major control of the pattern of visibility. At Great Dun Fell in the Pennines, which lies at an altitude of 848 m, the site is foggy or in cloud for about two-thirds of the time. Other curious high values are east of the Severn and south of London.

In recent years, because of a reduction in smoke emissions from industrial and domestic sources, the amount of poor visibility experienced in polluted areas has declined greatly. Likewise, the number of hours of sunshine have increased. For example, in Oxford, since the early 1960s fogs are now about half as frequent as they were in earlier decades.

Figure 76 is a map of the spatial variability in fog over Britain, which shows how the spatial pattern has changed between 1950 and 1983. Rural and coastal stations have tended to demonstrate a generally low and consistent frequency of fog over the period (see e.g. Tiree, Ronaldsway, Valley, Aberporth, St Mawgan, Manston, Leuchars, Dyce, and Wick). By contrast, at the early part of the period, those stations in close proximity to centres of industry (e.g. Abbotsinch in the Central Lowlands of Scotland, Tynemouth in the north-east, and Elmdon and Watnall in the Midlands) have very high fog incidences, but these high levels have declined sharply as a response to improving air quality resulting from Clean Air Legislation.

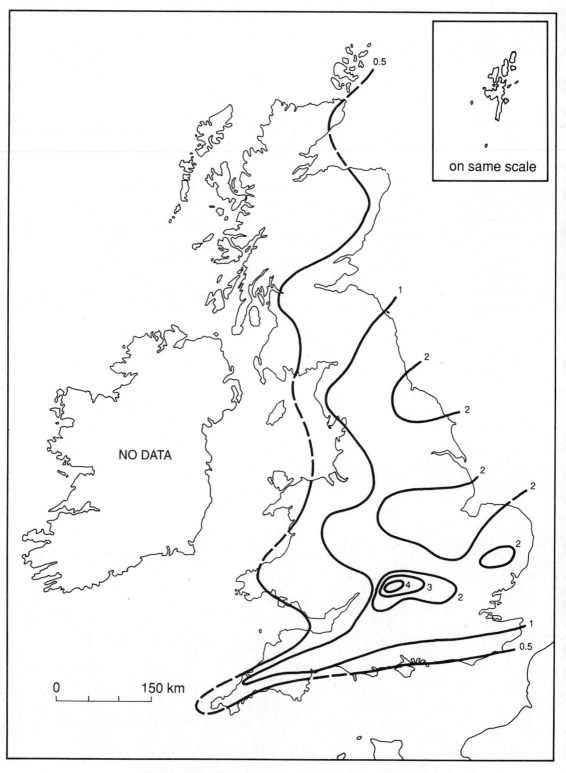

Fig. 75. The percentage of time with visibility less than 200 m based on observations at 03.00, 09.00, 15.00, and 21.00 hrs for the period 1962–71.

(*Source*: UK Meteorological data.)

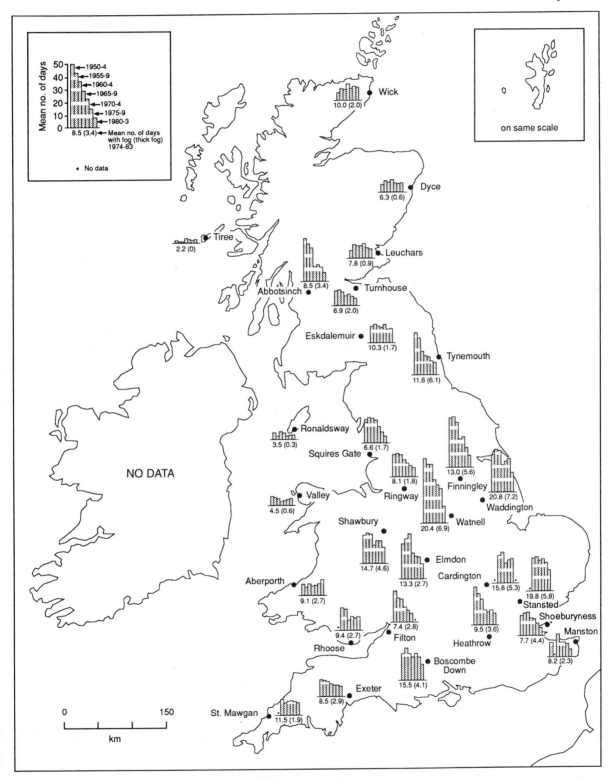

Fig. 76. The spatial variation of fog for the period 1950–83.

(*Source*: Musk 1991, fig. 6.6.)

Air temperatures

Figures 77 and 78 must be the most average maps ever produced. They are maps of averages of daily mean temperature for January and July respectively. The averages refer to the period 1941–70, and the average daily mean temperature is defined as half the sum of the average daily maximum and minimum temperatures. The values are corrected to mean sea-level values, so that to obtain local values it is necessary to substract 0.6 °C for every 100 m increase in altitude.

The January map, Figure 77, demonstrates that in winter the isotherms run mainly from north to south, so that temperature is lower in the east of the country than in the west. The pattern is heavily influenced by the modifying effect of warm sea temperatures. Thus, winter mean temperatures in an area like the East Midlands of England are comparable to those of the Orkneys and Shetlands!

The July map, Figure 78, shows a rather different trend. There is a marked latitudinal gradient of temperature from around 17 °C in south-east England to about 14 °C in north-west Scotland as the higher received solar radiation overrides the ocean effect.

Extremes of air temperature confirm these average descriptions (Figure 80).

Figure 79 shows the average minimum daily temperatures for January again reduced to sea-level. The lowest temperatures are recorded for the most part in inland areas in the eastern parts of the country (e.g. the Weald, the interior of East Anglia, and the uplands of Scotland). Parts of north Wales also experience low average daily minima in January due to altitude.

Figure 80 shows the average maximum daily temperature for July reduced to sea-level. Values range from about 14.5 and 15 °C in the far north and west of Scotland to 22.5 °C near London. The moderating influence of the sea is again evident, with a steep inland temperature gradient. A nice complementary map is that of the number of months with a mean temperature of more than 6 °C for the growing season (Figure 81). Only the northern uplands have more than six months with low temperatures. The west and south coasts have 9–12 months of comparative warmth. The rest of the country is uniformly mild.

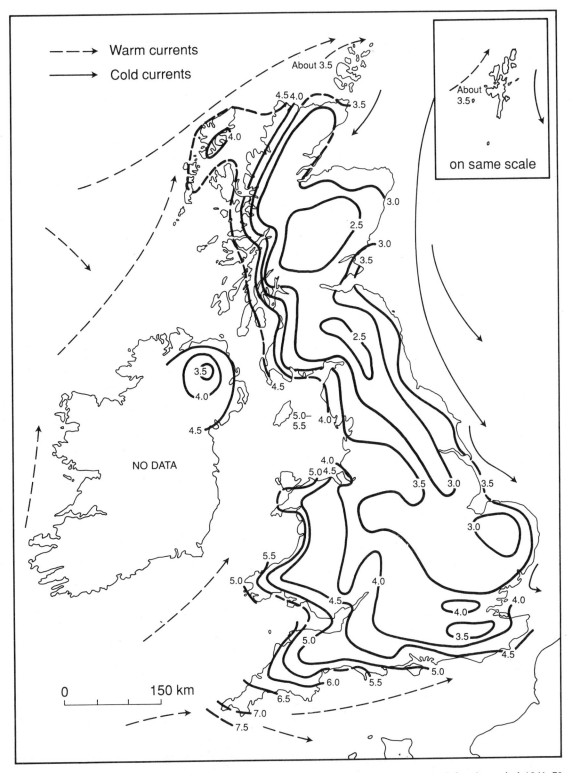

Fig. 77. Mean monthly air temperatures in January averaged for the period 1941–70 in °C and reduced to mean sea-level. To obtain local values subtract 0.6 °C for every 100 m of height.

(*Source*: UK Meteorological data.)

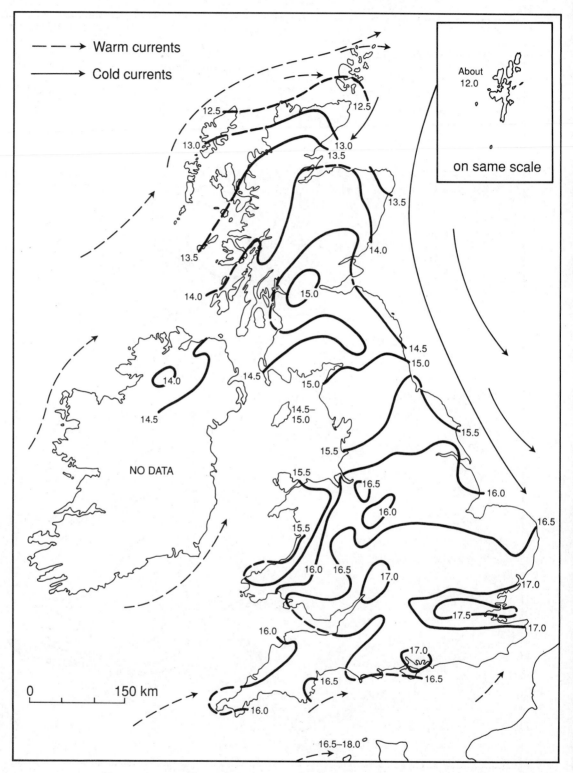

Fig. 78. Mean monthly air temperatures in July averaged for the period 1941–70 in °C and reduced to mean sea-level. To obtain local values subtract 0.6 °C for every 100 m of height.

(*Source*: UK Meteorological data.)

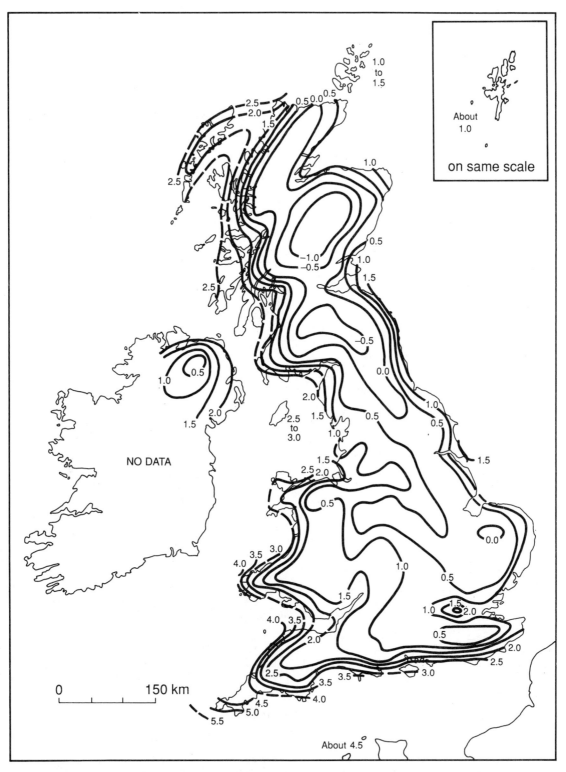

Fig. 79. Average minimum daily temperatures (°C) for the period 1941–70, reduced to mean sea-level.

(*Source*: UK Meteorological data.)

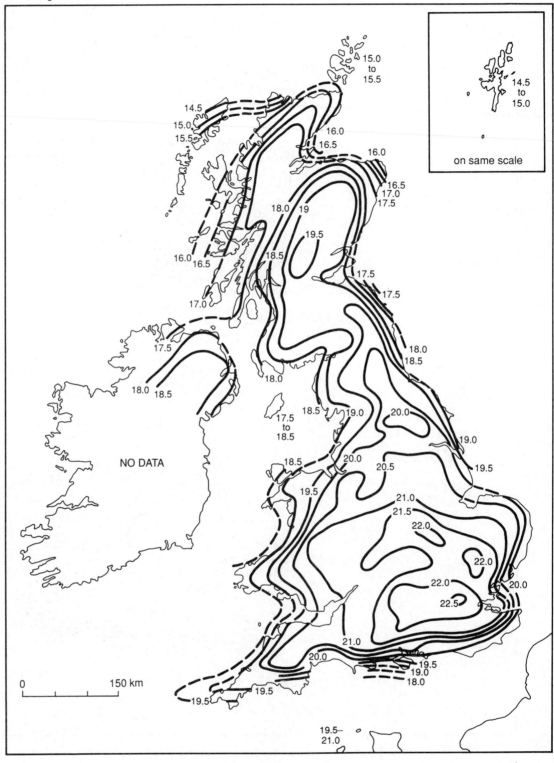

Fig. 80. Average maximum daily temperatures in July (averaged for 1941–70), in °C and reduced to mean sea-level.

(*Source*: UK Meteorological data.)

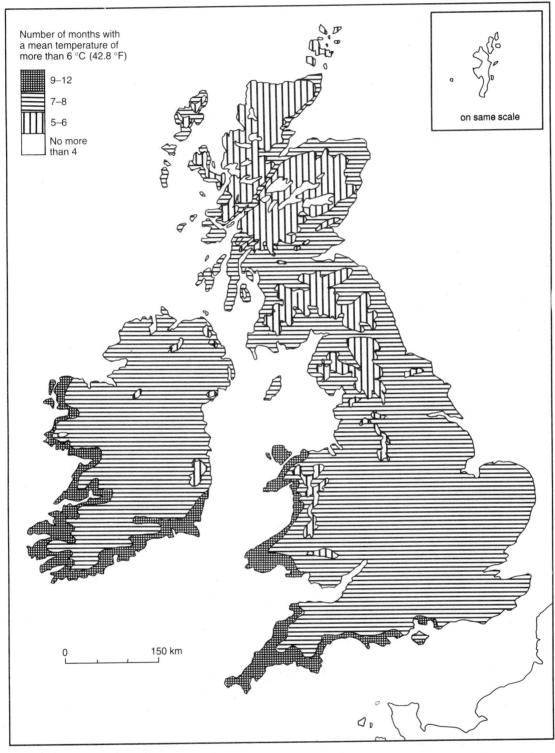

Fig. 81. The length of the growing season defined by the number of months with a mean temperature above 6 °C.

(*Source*: Gregory 1964, fig. 4.)

The 1990 heatwave

During the first four days of August 1990, there was a record-breaking heatwave of short duration over the British Isles (Figure 82). On 3 August, Cheltenham in Gloucestershire experienced a temperature of 37.1 °C, the highest recorded temperature measured in a Stevenson Screen in the UK. Temperatures in excess of 35 °C occurred over a large area of England and Wales. The highest temperature prior to this event was 36.7 °C, which was recorded at several stations in central and southern England in August 1911.

The conditions required for such hot spells are (apart from the season being the summer) warm conditions over the Continent, and cloud-free anticyclonic conditions which result in an easterly or south-easterly airflow, drawing hot air off the continent. It also helps if the airflow has only a short journey across sea, and if the ground is already dry as a result of preceding drought conditions.

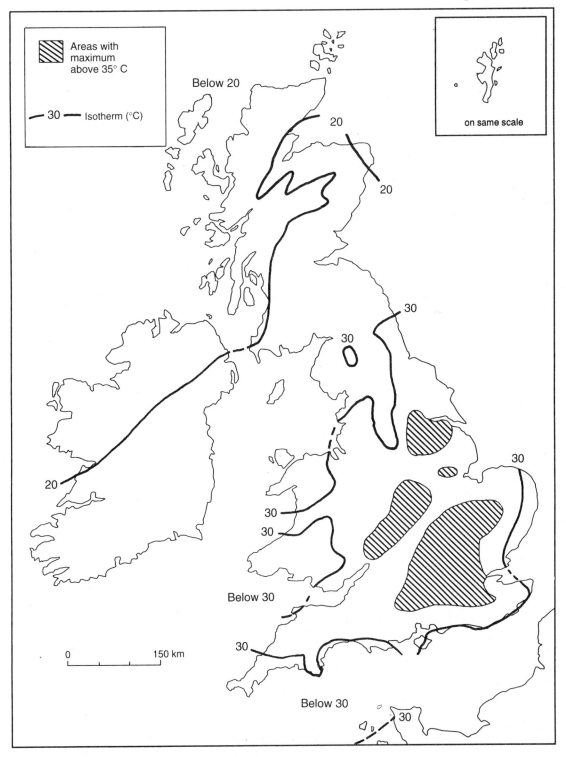

Fig. 82. Maximum temperatures observed on 3 August 1990. Isotherms are drawn for 20, 30, and 35 °C. Shading denotes regions with maxima above 35 °C.

(*Source*: Brugge 1991.)

Frosts

Frosts are climatic phenomena feared by horticultural-ists and motorists alike. They show great local variability in occurrence according to local topographic features. Some localities are notorious frost hollows. Figure 83 cannot take account of all local extremes, but it does give an indication of the patterns over the country as a whole. It shows the average annual number of days with a minimum air temperature of 0 °C or below for a 10-year period extending from 1956 to 1965.

Frosts are comparatively rare in the equable west of the country. Thus over a 10-year period there were less than 20 frosts per year in parts of Cornwall and Pembrokeshire, and even in the far north-west of Scotland, there were 30 days or less with frost in the year. The greatest frequency of frosts is in the Scottish Highlands, where the average number of days with frost in the year can exceed 120 (roughly one day in three). Large areas of south-east England, away from the immediate vicinity of the sea, have more than 60 frosty days per year. Rather startling is the high frequency (>70 days) in the Weald and Salisbury Plain!

Figure 84 gives another indication of the distribution pattern of conditions of extreme cold. In this case, the map shows the annual minimum temperature likely to occur once in 50 years at sea-level. The values range from about –6 °C in the Isles of Scilly, to –18 °C in extensive tracts of East Anglia, the Midlands, and the Weald, to more than –20 °C in the Highland regions of Scotland.

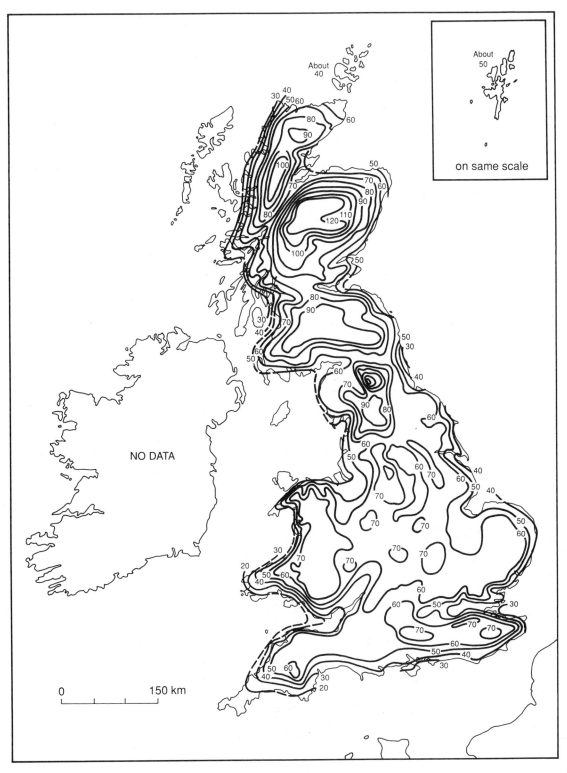

Fig. 83. Average annual number of days with minimum temperatures of 0 °C or less for the period 1956–65.

(*Source*: UK Meteorological data.)

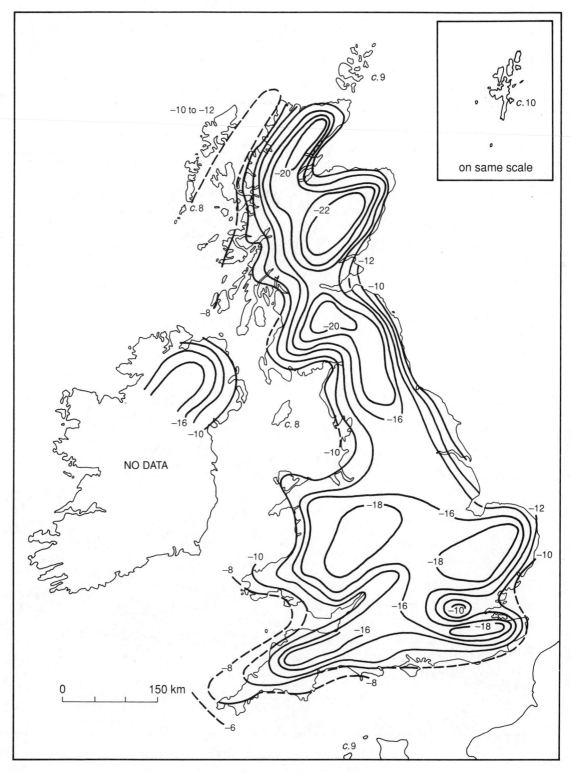

Fig. 84. 50-year extreme minimum temperatures.

(After Hopkins and Whyte 1975, in Perry 1981, fig. 3.3.)

Thunder

Thunder is difficult to observe and map, so that there may be inaccuracies in the patterns that are shown. Thunderstorms, defined as combinations of thunder and lightning with or without precipitation, are most frequent in the south and east of England (Figures 85 and 86). There is also a tendency for higher incidences to occur over land rather than over water, and over high rather than low relief.

Many of the thunderstorms of south-east England, where there may be 15 to 21 days with thunderstorms in the course of a year (Figure 87), develop locally, or originate on the continent of Europe. The thunderstorms of highland Britain (where there may be as few as 3 or less thunderstorms in the course of a year) tend to be more closely associated with cold fronts and showery west to north-west airstreams. The patterns should be compared with that of big storms (Figure 46).

One particularly intriguing feature of the map of thunderstorm activity is the strong peak over Greater London. There is reason to believe that the city itself plays a role in generating this peak. Detailed analyses have shown a very striking similarity in the morphology of the thunderstorm isopleths, and the urban area. Moreover, other workers have shown that the incidence of thunderstorms has increased as the urban area has increased. The mechanisms that may be involved are complex, but probably include higher temperatures and thermal convection as a result of the 'urban heat-island' effect, increased vertical motions from mechanically-induced turbulence over large buildings, and the obstruction of flow over the city.

Fig. 85. The frequency of thunder expressed in terms of the mean annual number of days when thunder was heard for the period 1901–30.

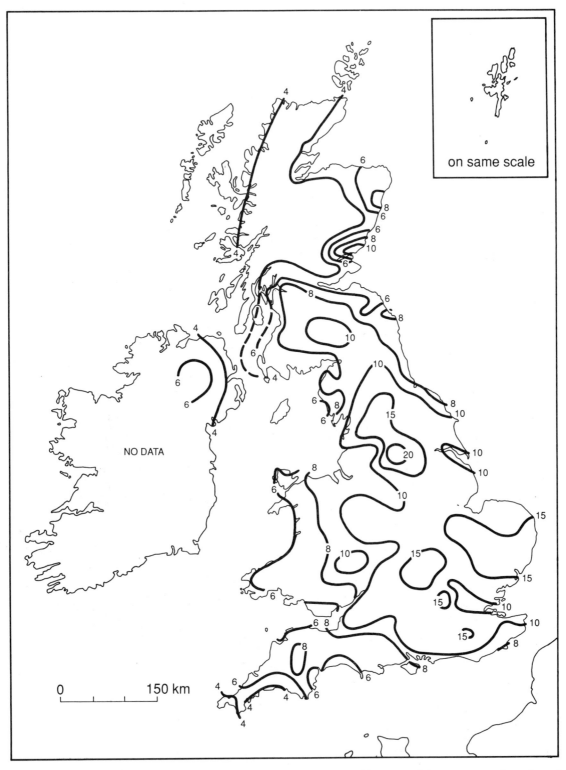

Fig. 86. The frequency of thunder expressed in terms of the mean annual number of days when thunder was heard for the period 1931–60.

(*Source*: UK Meteorological data.)

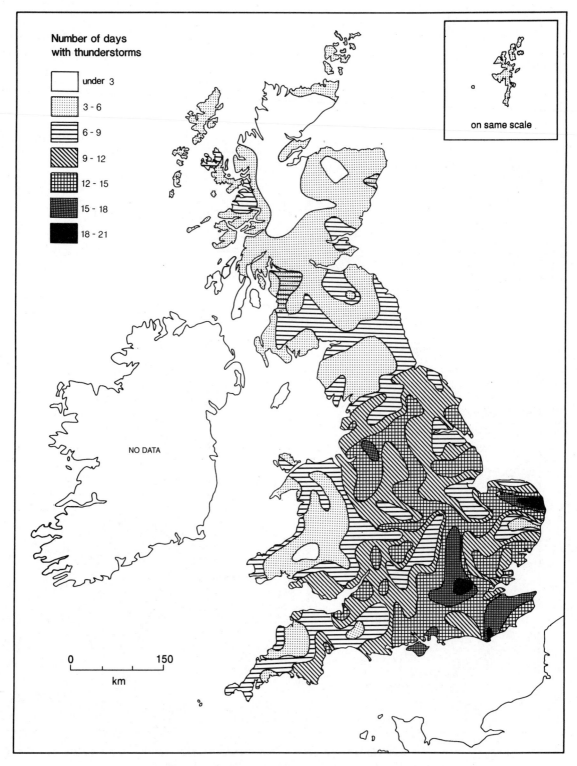

Fig. 87. Distribution of days with thunderstorms for the period 1955–64.
(Source: Perry 1981, fig. 2.11.)

Hydrology

Some river characteristics

One way of describing the nature of river networks is to obtain a measure of the length of drainage channel per unit area. Such a measure is termed 'drainage density'. In areas of intense erosion, such as badlands, one would expect the values to be very high, whereas in areas where runoff and fluvial erosion are limited one would expect values to be low. In general, and despite a reputation for frequent rainfall, drainage density values in Britain are low by world standards, and are less than about 0.3 km of channel per km^2 over much of the country (Figure 88a). It is only in the north and west that there are extensive areas where drainage density values are greater. This pattern reflects impermeable rock type as much as rainfall.

Stream frequency, expressed as the number of streams per unit area, provides an alternative measure of river network density in the landscape. Regional patterns are even clearer than they are for drainage density. Note the comparative sparseness of the network over lowland Britain and the relatively high values of the western uplands in Scotland, north Wales, and the south-west of England (Figure 88b). Values range from a minimum 0.007 streams per km^2 to a maximum of 0.107. Among the reasons for the low values of drainage density and stream frequency in Britain are the generally low rainfall intensities, the widespread development of woodland, grassland, and cropland, and the high infiltration capacities of most soils.

A further way of describing the landscape is to measure the number of lakes per unit area. Here again some marked patterns become evident. Whereas the average frequency of lakes is about 0.024 per km^2, much higher than average values occur in parts of Scotland (particularly the west), the Lake District, Lancashire, and north-west Wales (Figure 88c). It is interesting to note that these areas broadly correspond to zones of maximum intensity of glacial erosion (see Figure 24). The highest values of all (c.0.4 per km^2) occur in the heavily scoured landscape of the Hebrides. Not all lakes are, however, the result of glacial erosion. In the Lancashire–Cheshire–Shropshire lowlands, some depressions may be the result of surface irregularities caused by the wastage of glacial deposits or subsidence, while in other parts of the country (e.g. Norfolk) depressions may result from a combination of processes including periglacial activity, solution of the chalk, and various types of human excavation including marl pits and peat workings (e.g. the Broads).

Urbanization of previously rural catchments is another important hydrological characteristic, for urbanization changes stream behaviour through the reduction in vegetation cover, evapotranspiration, and soil infiltration capacity. Flood risk may be increased and water quality reduced. Values of the percentage of surface area under urban development are, as might be expected, high in the Central Lowlands of Scotland, in south Wales, and in the industrial heartland of England (Figure 88d).

Fig. 88. Some river characteristics. (a) Drainage density; (b) stream frequency; (c) Lake frequency; (d) Percentage of area under urban development.

(*Source*: Lewin 1981, figs. 1.8, and 1.9.)

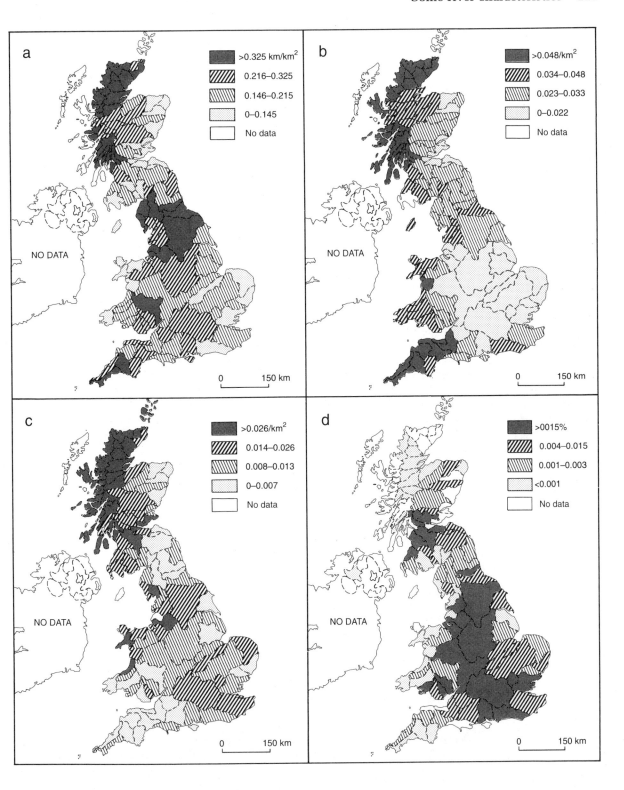

Flow patterns

Most British rivers show a relatively simple pattern of flow during the year, with a single period of high flow (in the winter) followed by a period of low flow (in the summer). However, there are some interesting patterns in the timings of discharge maxima and minima across the country (Figure 89a and b), for despite some differences in detail it is clear that the two distributions are broadly similar in that the time of occurrence of both largest and smallest flow becomes gradually later towards the south and east. The mean maximum flow month in Scotland, north-east England, and western Wales occurs in December, but as late as February or March in parts of the east. Similarly, the mean minimum monthly flow occurs as early as May in parts of Cheshire, in June over central and western Scotland, northern England, and portions of north and west Wales, and as late as September in parts of eastern and central-southern England.

The reasons for this pattern are complex. Part of the explanation may lie in the seasonality of precipitation. The earlier occurrence of periods of low flow in the west and north may result from the fact that average monthly rainfalls decline early in the year in the mountainous north and west. Even more important, however, may be the increasing effectiveness of evapotranspiration losses towards the south and east, and the fact that the sedimentary rocks of lowland Britain have a great groundwater storage capacity, which encourages infiltration, percolation, and gradual release into stream channels.

The quantity of stream-flow also demonstrates a clear regional pattern related to the pattern of precipitation and evapotranspiration. Figure 89c shows mean annual runoff values (i.e. the residual of precipitation minus actual evapotranspiration). Low values (below 500 mm of precipitation equivalent) occur in the south-east of England, in the lee of the Welsh mountains, and along eastern coastal areas. High values, which may exceed 1500 mm, are confined to Dartmoor, the mountains of Wales and Cumbria, and the western Scottish Highlands. Figure 89d maps the discharge ratio (i.e. the proportion of precipitation that appears as river discharge). Values range from less than 25% in East Anglia to more than 75% in some western parts of Wales, Scotland, and northern Ireland.

Fig. 89. Flow patterns of British rivers. (a) The pattern of timing of the month of highest flow; (b) The timing of the month of lowest flow; (c) Mean annual runoff; (d) The discharge ratio. (*Source*: Lewin 1981, figs 1.11, and 1.15.)

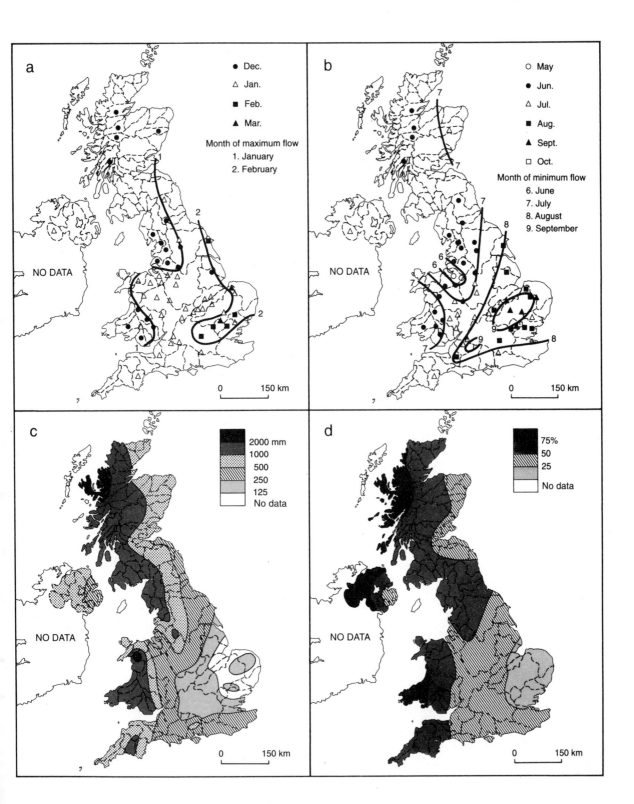

Streamflow

The pattern of mean annual streamflow reflects the familiar weather pattern of the high flow to the west and low to the east (Figure 90). Mean annual streamflow exceeds 1500 mm in the highlands of Britain. This reduces to less than 200 mm in the east.

The catchments of Britain (Table 4) are not large but their area is difficult to calculate because of the problem of defining estuaries as part of a catchment. The largest, based on the Humber estuary, is 25 000 km² incorporating the Trent and Ouse systems. The Thames estuary drains 15 000 km², the Severn–Avon system 11 500 km², followed in size by the Bedfordshire Ouse, Tay, Tweed, and Wye. It is noticeable that the area of many catchments extends across the W–E water-availability 'divide' and is more determined by the structural history of Britain.

The highest discharges are the Humber estuary (>200 m³ sec) and the Tay (160 m³ sec). Other rivers are quite small, certainly by world standards.

Table 4. The main rivers of Great Britain

River	Length (km)	Area (km²)	Mean annual discharge (m³ sec⁻¹)
Thames	239	9950	67.40
Trent	149	7490	82.21
Wye	225	4040	71.41
Tweed	140	4390	73.85
Devon	206	4330	62.70
Ouse	117	3320	40.45
Aire	114	1920	36.89
Tyne	89	2180	43.45
Eden	102	1370	31.02
Ribble	94	3030	31.72
Trent/Ouse	184	2210	14.16
Avon	125	2210	14.43
Tywi	82	1910	38.34
Tees	103	1260	19.46

Source: Ward 1981.

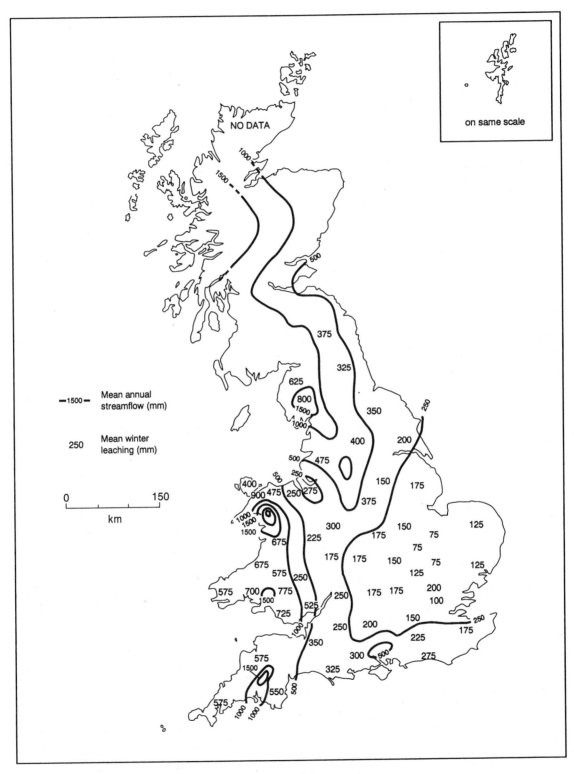

Fig. 90. Mean annual streamflow.

(*Source*: Chandler and Gregory 1976, fig. 7.6.)

Flooding

Flooding from either rainfall or snowmelt is a serious hazard in some parts of Britain, even though there is a long history of flood control and channel manipulation. Although the highest rainfall totals occur mainly in the higher areas of the west, the lowlands may from time to time have very high storm totals (Figure 46) that can generate appreciable floods.

In general the north and the west have the highest mean annual flood values (Figure 91a) and the highest maximum discharges (Figure 91b) per unit area. The former gives an index of flood potential in the form of an estimate of mean annual flood, while the latter gives an indication of flood experience during the period over which the river has been gauged in the form of the highest instantaneous gauged discharges. Mean annual flood values for lowland Britain tend to be below 0.25 m^3. sec^{-1} km^{-2}, whereas higher values, above 0.75, are restricted to Dartmoor, the Welsh mountains, areas of the Pennines and Cumbria, the Southern Uplands, and parts of western and northern Scotland.

In terms of flood risk to people, many different factors are involved in determining the hazardousness of any particular place or event. So, for example, the relief factor can be important, as at Lynmouth in August 1952, when 229 mm of rain fell in just 24 hours on a very steep catchment. Antecedent conditions are also important. At Lynmouth the catchment was saturated before the storm. By contrast the great storm that dumped over 280 mm of rainfall in a day on Martinstown in Dorset in 1955 had comparatively little effect, falling as it did on less steep terrain, underlain by chalk which was not saturated by previous rain. Floods can also be made serious in areas where there is sudden snowmelt (e.g. on rivers draining the eastern Pennines), where flood yields have been increased by the spread of urbanization (e.g. in some of the London suburbs), or where ponding occurs in low-lying areas (e.g. the Somerset Levels or the Fens). Some rivers have very low gradients, and the River Severn, which often floods in the vicinity of Tewkesbury, is at only 48 m above sea-level 130 km from the sea. Also highly significant in this respect is where people live. Thus although some Scottish catchments deliver high peak-flows they do little harm in that they pass through sparsely populated areas.

Table 5 presents some data on peak river flows for selected British rivers. It shows that there can be a 10 to 15 times difference between the mean flow of rivers and the highest daily mean. British rivers can be violent even if it is on a small scale.

Table 5. Examples of peak river flows in Great Britain

River gauge	Length of record (starting year)	Mean annual precipitation (mm)	Mean flow (m^3 sec^{-1})	Highest daily mean (m^3 sec^{-1})	Peak (m^3 sec^{-1})
Thames (Kingston)	1883	720	66.9	1059	—
Trent (Colwick)	1958	777	85.9	854.9	957
Tay (Ballathie)	1952	1422	158.8	1223	1570
Tweed (Norham)	1962	922	77.7	1138	1518
Severn (Bewdley)	1921	916	61.9	637	—
Wye (Redbrook)	1936	1023	71.8	—	905
Spey (Boat O'Brig)	1952	1103	64.3	1089	1675
Eden (Sheepmont)	1967	1185	50.8	772.9	1357
Dee (Woodend)	1929	1119	36.4	648.5	1133
Usk (Chain Bridge)	1957	1389	27.8	585.4	945

Source: Data in *Hydrological data, UK 1988 Yearbook* (IOH/NERC 1989).

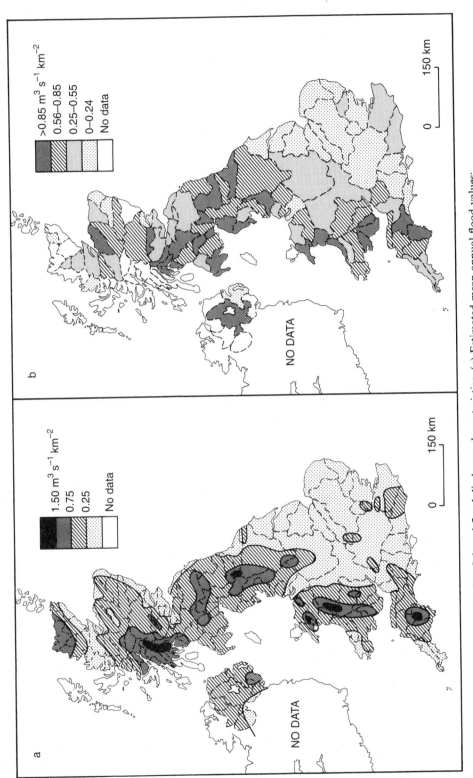

Fig. 91. Selected flood discharge characteristics. (*a*) Estimated mean annual flood values; (*b*) Mean maximum instantaneous gauged discharges for each hydrometric area.

(*Source:* Lewin 1981, fig. 1.18.)

Stream slope

British rivers are by world standards quite steep, particularly in the western (upland) areas. This is because the distance to the sea is short but with substantial relief. Only the Thames, Great Ouse, and Severn reach low, continental gradients (Figure 92a).

A characteristic of such streams is that they also exhibit high degrees of roughness. Their courses are dominated by pools, riffles, boulder stream courses, and a coarse bedload. The Pleistocene glaciation influence is strong with high volumes of coarse bedload as well as a high percentage of bare rock outcrops. Upper catchments are particularly steep and rough. The side slopes are also steep and often composed of stored, coarse-grained colluvium, while floodplains are small. This means that quite large events are needed to move the bedload. When these occur the changes are dramatic both to the channel and adjacent hillslopes.

Average mainstream gradients range from 0.2–30 m km^{-1} (Figure 92b). There is no consistent pattern at this scale over Britain although the highest gradients tend to occur on mountainous western coasts. Gradient is, however, important because it largely determines the rate of supply of gravitational energy at the channel bed. This is required to overcome friction, and low erosion and sediment transport. The useful way of measuring this is by using gross stream power measured in *watts* per unit length (Figure 92c) or specific power which is calculated and expressed (Figure 92d) per unit area of bed. The stream-power maps of Britain are remarkable. There is a wide range since steep slopes and high rainfall coincide. Again the familiar Highland–Lowland division becomes apparent.

Fig. 92. (a) Mountain streams shown by the morphometric variable stream slope; (b) The average gradient of major rivers; Stream power at bankfull discharge: (c) per unit channel length; (d) per unit channel area.

(*Source*: Lewin 1981, figs 4.2, 4.4, 3.2.)

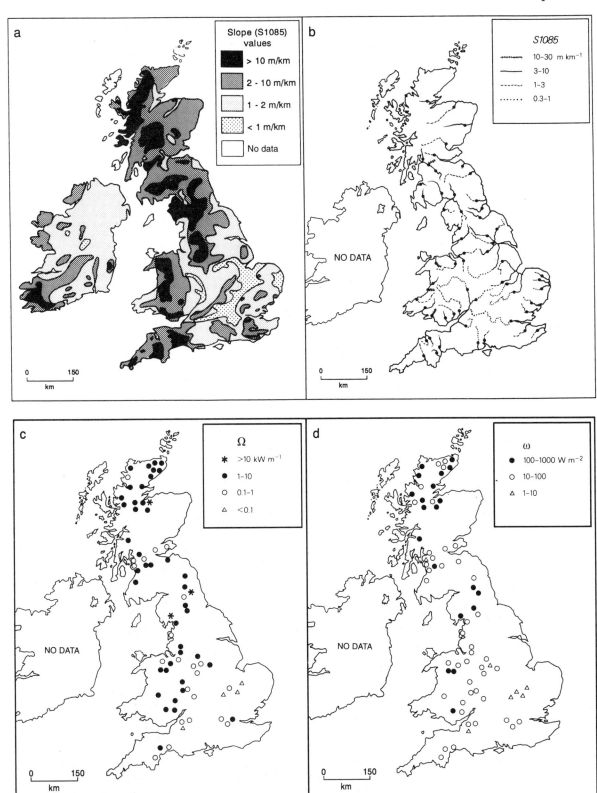

The chemical load of rivers

One way to try to quantify the pattern and importance of the denundation achieved by rivers in the British landscape is to undertake analyses of the various components of a river's load, be it dissolved, suspended, or rolled along the bed.

One method of estimating the amount of material dissolved in stream water is to determine specific conductance values using a conductivity meter (Figure 93a). The higher the specific conductance values the greater is the amount of material that is dissolved in the water. The map demonstrates the pattern of specific conductance values for Britain. The crucial importance of rock type (and associated elevation) is clear. Upland areas developed on resistant rocks (e.g. the Scottish Highlands, the Lake District, Snowdonia, mid-Wales, and parts of the south-west peninsula) have generally low specific conductance values, whereas the sedimentary, particularly the calcareous rocks of lowland England, to the east of the Tees–Exe line, have much higher values. It is also necessary to understand the dilution effect of precipitation amount if this pattern is to be explained. Not only do the resistant rocks generate less dissolved material than the more susceptible, younger sedimentaries, but being upland areas they also have high precipitation levels and so generate large amounts of runoff. The actual total volume moved can therefore be high even though the concentration is weak.

The pattern of total removal of dissolved material in tonnes km^{-2} y^{-1}, involves such a consideration (Figure 93b). Areas in the east of England have low or medium dissolved solid ratings (because of the low amount of runoff) as do upland areas (which have highly resistant

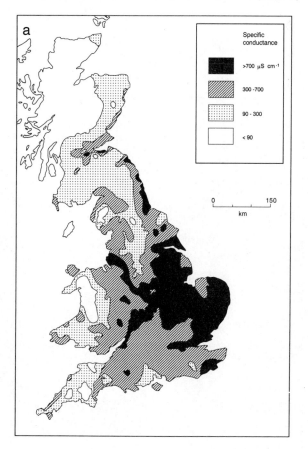
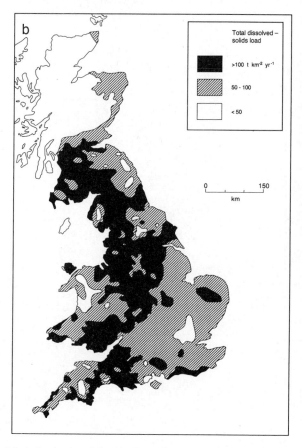

Fig. 93. Soluble loads. (*a*) Mean background loads of specific conductance in streams based primarily on data collected in the period 1977–9. (*b*) The pattern of annual total dissolved solids in streams.

(*Source*: Walling and Webb, figs 5.3, and 5.9.)

rocks). Areas in between these two extremes tend to have the highest values.

If allowance is to be made for the materials that may be brought into stream catchments in precipitation then it is possible to subtract these from the total dissolved load figures to obtain an estimate of rates of chemical denudation. For most of Britain rates are relatively low, with most values being equivalent to a denudation rate of less than 40 m^3 km^{-2} y^{-1} (Figure 94).

Data for estimating rates of denudation associated with suspended transport are much less readily available, reliable, or widespread. There appears, however, to be a very wide range in values, between 1.0 and almost 500 tonnes km^{-2} y^{-1}. Probably a value of around 50 is fairly typical. By world standards this is low, and results from low rainfall intensities, the relatively dense vegetation and crop cover, and good land management. One or two other generalizations may be possible (Walling and Webb 1981: 167):

Loads in excess of 100y km^{-2} y^{-1} would seem to be associated primarily with upland areas receiving annual precipitation greater than 1000 mm and, within these areas, with small- and intermediate-sized catchments where sediment delivery ratios will be relatively high. Conversely low suspended sediment yields (<25 t km^{-2} y^{-1}) would appear to reflect low annual precipitation (e.g. River Welland and River Nene), large basins where sediment delivery ratios will be relatively low . . . and low relief . . . Very low suspended sediment yields (<5 t km^{-2} y^{-1}), represented by the East Twin catchment on the Mendip Hills and the Ebyr N. and Ebyr S. catchments in central Wales, may be accounted for in terms of the small headwater areas involved, the resistant bedrock and the essentially undisturbed conditions found in these upland areas.

The map of chemical denudation (Figure 94) does reflect

the W–E division of Britain that is based on relief and rainfall, but there are significant differences. Solute loads reflect the presence of soluble rocks in the catchment as well as the application of fertilizers and other industrial effects. Even this pattern is not clear, however, because soluble rocks themselves do not yield solutes if the rainfall conditions are not suitable. Thus, on the map, the Chilterns chalk escarpment is not visible although the South Downs, Salisbury Plain, Dorset chalklands, and the Cotswolds are. The carboniferous limestone of south Wales and the Peak–Pennine area are obvious centres of chemical denudation.

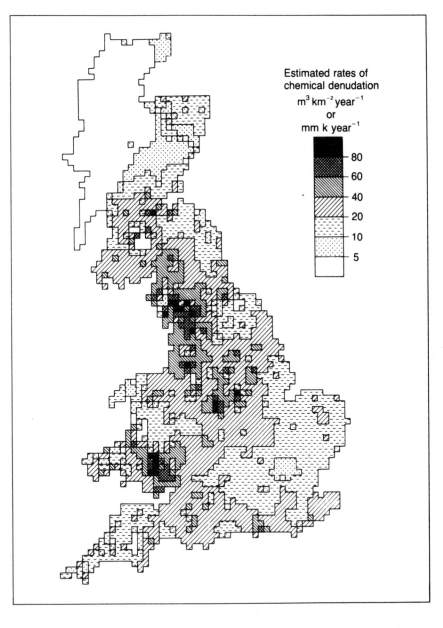

Fig. 94. Estimated rates of chemical denudation.

(*Source*: Walling and Webb 1986, fig. 7.18.)

Suspended sediment

It is not possible to give a countrywide pattern of suspended sediment transport in British rivers because the measurement of this variable is not included in most current monitoring programmes. This is largely because suspended sediment is not generally regarded as a problem. Reservoir sedimentation is minimal in most areas and does not appear to present difficulties for water treatment. Soil erosion is not really a serious problem in world terms and channel sedimentation does not cause water flow difficulties.

Although there have been attempts by academic researchers to establish the rates of denudation most of the measurements face the difficulty of short-term records, manual sampling, and restricted results from automatic turbidity-sensing measurement programmes. The data rarely exhibit contemporaneity and so maps of mean annual sediment yield must be treated with care (Figure 95).

A further difficulty is that the methods used to calculate sediment yield from a catchment, especially if based on proxy variables or equations, yield non-comparable results. If the sampling programme is of coarse scale it will probably miss the main sediment transport events which are represented by just a few floods.

Bearing these difficulties in mind the following generalizations are made. First, the maximum levels of British rivers rarely exceed 500 mg l^{-1} and then only following severe human activity, forest ditching, housing development, and road construction. Rarely, such as from spillage or china-clay workings, values as high as 50 000 mg l^{-1} have occurred, but these are unfortunate errors, rapidly corrected. Put in spatial terms the yields are from 1–500 t km^{-2} y^1 with an average of 50 t km^{-2} y^1. This is low in global terms.

Secondly, there is not much variation between rivers, no matter what part of the country is examined. There is actually as much range between winter and summer on one river as there is between different regions.

Thirdly, the main size of sediment transported is silt and clay. Organic matter content ranges from as low as 2% from upland rocky catchments to 60% on lowland floodplains. Averages are in the order of 20% for most regions (10–30% range). This percentage tends to increase in the summer months when productivity is high.

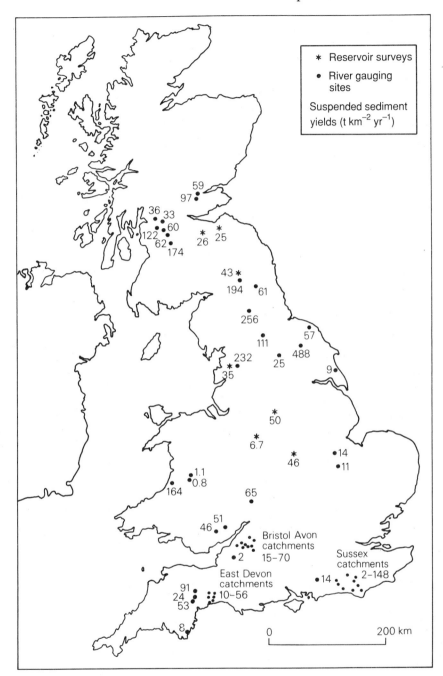

Fig. 95. Suspended sediment yield reported from river-gauging stations and reservoir surveys.

(From Lewin 1981, fig. 5.22.)

Chloride concentrations in rivers

The chloride ion is generally considered to be an atmospheric input brought by precipitation from ocean sources. Thus Britain, surrounded by sea and subject to westerly winds from the Atlantic receives considerable quantities of sodium chloride. There are other groundwater sources but these are local.

The map therefore shows quite distinct patterns (Figure 96). The source is obviously related to the sea, to the prevailing west and south-west winds, and to the amount of precipitation which is higher on the western uplands. The solute concentrations in rain range from 5 to 35 mg l^{-1} and are highest in the west. Despite this the concentrations in rivers are highest in the east. This is because in the west there are high precipitation totals, low evapotranspiration, and, therefore, a high runoff: recipitation ratio. In the east this ratio decreases sharply and so the concentrations are not so diluted. The actual concentrations increase from 10–30 mg l^{-1} in the west to 60 mg l^{-1} in the Suffolk, Norfolk, and Essex runoff.

Two other patterns are obvious. Most coastal areas have higher loadings than their hinterlands due to the proximity of the sea. However, anomalously high concentrations occur where there is an obvious geological source of chloride. The Cheshire Basin, for example, derives chloride from the Triassic salt strata. Large urban areas may also cause higher than normal contributions due to domestic, industrial, and highway use.

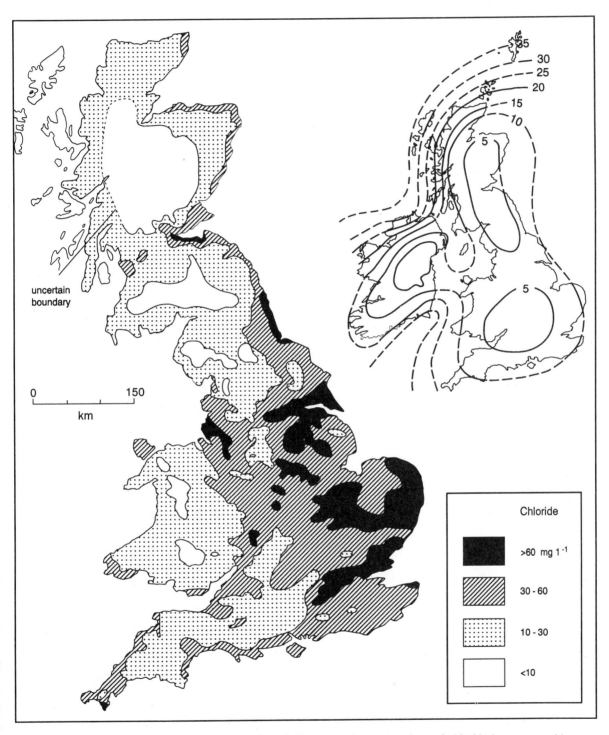

uncertain
boundary

0 ———————— 150
km

Chloride

■ >60 mg 1⁻¹

▨ 30 - 60

⠿ 10 - 30

□ <10

Fig. 96. Variations in background concentrations of chloride in streams with mean concentrations obtained from data primarily collected in the period 1977–9. Inset depicts annual chloride medians of rainfall over Britain.

(*Source*: Lewin 1981, 5.5.)

Nitrate loads in British rivers

Recent years have witnessed growing concern over the trend of increasing nitrate concentrations in surface and ground waters. Fears have been expressed that if such increases are occurring then they could cause eutrophic degradation of rivers and lakes, and create health risks for humans. Major reasons for increasing nitrate concentrations include the use of synthetic fertilizers, the application of sewage sludge to fields, the release of sewage into streams, and the ploughing-up of soils.

Mean annual nitrate (NO_3–N) concentrations in

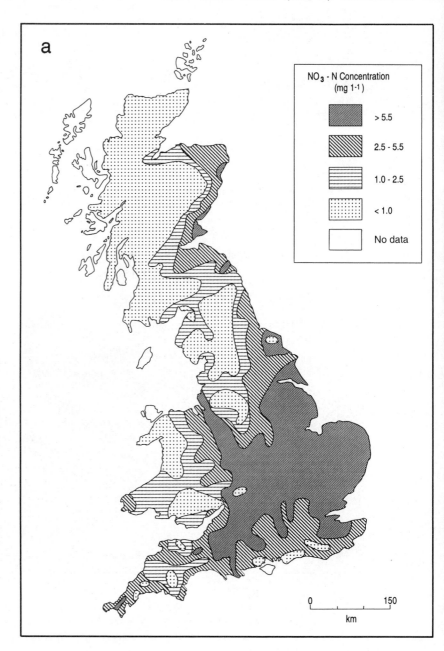

Fig. 97. Mean annual nitrate concentrations in British rivers expressed as NO_3–N in mg per litre for (*a*) the present time, and (*b*) for the year 2000.

(*Source*: Betton *et al.* 1991.)

rivers have been mapped in Figure 97a. They range from 0.1 to 15 mg l⁻¹, and a marked north-west to south-east gradient is evident. This gradient probably reflects three main factors: relief, hydrometeorological conditions, and agricultural activity. Thus the uplands of the north and west generally have low mean concentrations (<1.0 mg l⁻¹) because precipitation tends to dilute rivers' solute concentrations. There is also limited agricultural activity. By contrast, many lowland areas have mean NO_3–N concentrations that range from 5 to 9 mg l⁻¹, and in the drier, intensively cultivated regions of East Anglia values may be over 10.

Analysis of large numbers of basins in the UK has revealed that in recent years NO_3–N concentrations have significantly increased. It is predicted that over the next decade (Figure 97b) there could be an average increase of from 2 to 3 mg l⁻¹ for NO_3–N concentrations. This would mean that much of eastern England will then have mean concentrations that will exceed the European Community's acceptable level of 11.3 mg l⁻¹.

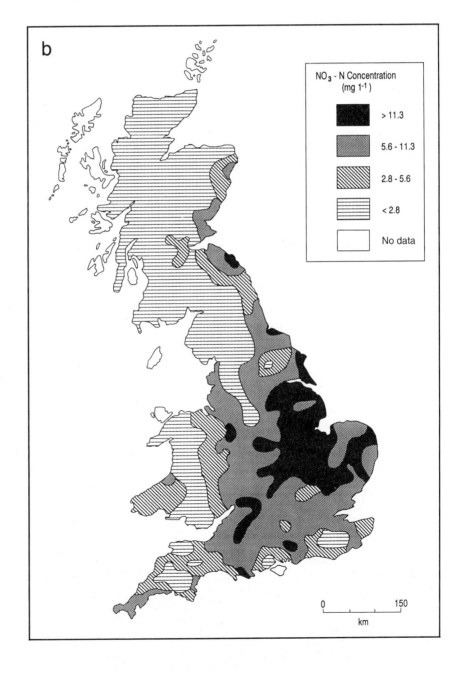

Biogeography and Soils

The colonization of the landscape by trees

In the late Pleistocene, much of western and northern Britain was overlain by a great ice-cap, and southern England was for the most part a tundra landscape, where trees were, in all probability, largely absent. However, as the climate warmed up during the Holocene or post-glacial times, trees started to return to the British landscape. By about 5000 BP* (see Figure 99), all but the very highest and most exposed portions of the British Isles were forested. Figure 98 shows the progress of tree migration in the early Holocene. It is based on an interpretation of the pollen records from a large number of sites that have been dated by radio-carbon.

The pattern of colonization is different for different species. Birch (*Betula*), which is a rapid colonizer of areas that have recently been vacated by ice and is a dominant species in many marginal areas in northern Europe today, moved very rapidly into the British Isles between 10 000 and 9500 BP, spreading westwards from the North Sea basin. Another fast early colonizer was hazel (*Corylus*) (Figure 98a), but this tended to spread eastwards from the Irish Sea basin. Its rate and spread of colonization was very high at first (500 m per year). Elm (*Ulmus*) first colonized south-east England at c.9500 BP and then spread northwards and eastwards. Oak (*Quercus*) on the other hand seems to have made its entry from the south-west (Figure 98b). Pine (*Pinus*) spread out from three main areas (south-west Ireland, southern England, and north-west Scotland) (Figure 98c). Lime (*Tilia*) came in rather later, colonizing south-east England at about 7500 BP (Figure 98d), while the beech (*Fagus*) was even later, reaching the south-east of England only by 3000 BP (Figure 98e).

* Years before present.

Fig. 98. Some examples of the spread of various tree types. (After Birks 1990): (a) Spreading of *Corylus avellana* (hazel). The contours, in years before present (BP), show its likely range limit at 500-year intervals. The shaded area is where hazel was present prior to 9500 BP. (b) Spreading of *Quercus* (oak). The contours, in years BP, show its likely range limits at 500-year intervals. Note the slowing in its spread after 8000 BP. (c) Spreading of *Pinus sylvestris* (pine). The contours, in years BP, show its likely range limits at 500-year intervals in England and Wales and in Ireland and at 1000-year intervals in Scotland. (d) Spreading of *Tilia* (lime). The contours, in years BP, show its likely range limits at 500-year intervals. Note the slowing in its spread after 7000 BP. (e) Spreading of *Fagus sylvatica* (beech). The contours, in years BP, show its likely range limits at 500-year intervals. Note the absence of any slowing in its spread.

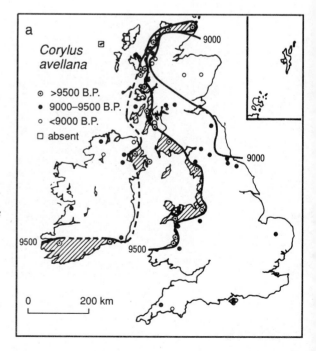

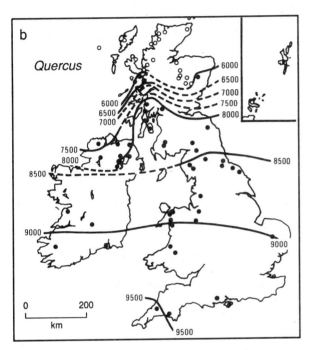

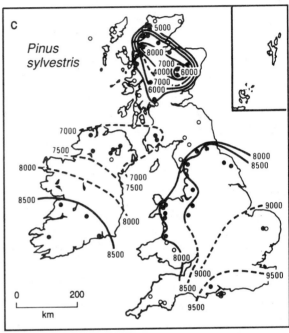

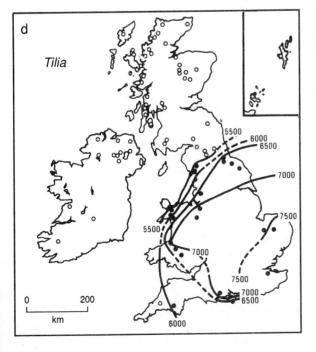

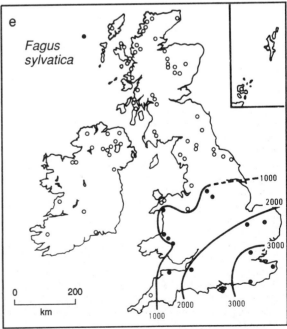

Forest types 5000 BP

After *c.*5000 BP, humans began to disrupt the pattern of vegetation in the British Isles in a substantial way. Human population levels were growing, and the adoption of the plough was associated with an expansion of farming activities. 5000 BP probably represents the time of most complete development of British and Irish woodlands before clearances began to break up the pattern. It is thus of some interest to see what the pattern of forest cover was at that time.

Figure 99 is an attempt to reconstruct that pattern. It is based on several different methods of data analysis and generation. Pollen analyses were used to identify the tree species that were present, and this information was combined with what is known of the relationship between particular species and particular soil types.

The number of tree species was not large. The main trees were two species of birch (*Betula*), pine (*Pinus sylvestris*), willow (*Salix*), hazel (*Corylus avellana*), oak (*Quercus*), elm (*Ulmus*), alder (*Alnus*), ash (*Fraxinus*), lime (*Tilia*), and yew (*Taxus*). Beech (*Fagus sylvatica*) and hornbeam (*Carpinus betulus*) had not reached Britain by this time, but are, of course, important species today. Beech arrived at around 3000 BP (Figure 98) and hornbeam at around 1000 BP.

Certain interesting patterns emerge. First, most of the country was covered by forest of one type or another, the only exceptions being very exposed or high altitude situations. Pine was important in western Ireland and in the Highlands of Scotland, but was relatively rare elsewhere. Deciduous species were predominant. Hazel was particularly important in the central portion of Ireland, where calcareous soils are widespread, while oak was more prevalent in the east.

In Scotland, birch forest was present over large areas of the wet, windy, and bleak north-west, where it did not suffer from competition from other species, while oak was significant in the lower altitude areas to the south and east. Oak was also widespread in Wales, except in the higher areas, and in the south-west peninsula. In central and southern England lime was dominant on fertile, non-calcareous soils. It is likely that, on the chalk and other limestones, ash was dominant.

In Ireland, the pattern is broadly similar with hazel dominating the interior and oak the south and north. The unusual feature is the cover of pine forest on all the western uplands and coastlands, including the marine islets of the south-west.

Fig. 99. Provisional map of the predominant woodland types for the British Isles 6000 BP.

(*Source*: Bennett 1989, fig. 1.)

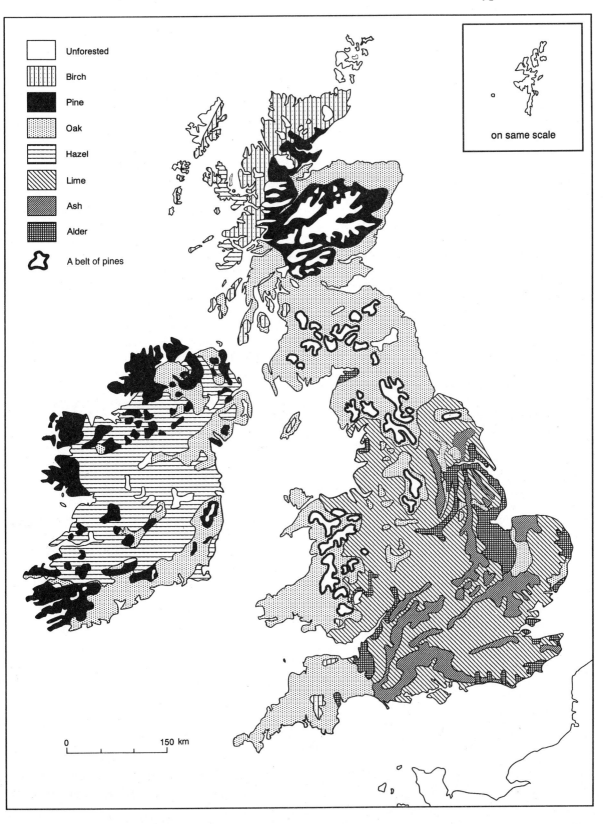

Unforested

Birch

Pine

Oak

Hazel

Lime

Ash

Alder

A belt of pines

on same scale

0   150 km

Soils

Soils display great variety in their characteristics according to the range of factors that control their development. Important controlling factors include climate, parent material, vegetation cover, topographic position (or relief), and age. In the classification used as the basis of this map, prominence is given to the role of soil drainage, with four categories of poorly drained soils and four categories of well-drained soils (Figure 100).

Of the poorly drained soils mention must first be made of acid-upland peats and peat bogs. These are a feature of upland areas in the higher, wetter parts of western Britain, where severe leaching takes place, and where waterlogging is promoted. Moreover, the lower temperatures of such areas lead to a reduction of biological activity and a slower breakdown of plant debris by biological reducers. A second type of peat soil—lowland fen peat—occurs in waterlogged areas at lower altitudes such as the Fenlands of East Anglia and the Somerset Levels. These peats are rather more alkaline than the first category.

Also within the poorly drained soils category are two types of gley soils. Their profiles reflect the role of periodic waterlogging. In one case the waterlogging is caused by being in a topographic situation where a groundwater table approaches the surface (e.g. low-lying areas subject to flooding by the sea or by rivers). In the other it is caused by the presence of an impermeable substratum of parent material as, for example, in the extensive clay vales of southern England. Gley soils display grey horizons with ochreous mottling which develops where soil pores are filled by water containing dissolved organic substances, which, with anaerobic bacteria, bring about the reduction and solution or iron compounds.

With respect to well-drained soils two types of brown earth are of great importance. These are sometimes termed brown forest soils. They are freely or moderately well drained, and contain no free calcium carbonate, or redistributed organic matter, iron, or aluminium. The colour tends to be more or less uniform throughout the profile, though under forested conditions they are usually characterized by a surface accumulation of leaf litter, underlain by mull humus.* Argillic brown earths display a subsurface horizon of clay accumulation consisting of clay translocated from higher up the profile. They frequently develop on areas with mudstone as a parent material.

Podzolic soils, on the other hand, develop preferentially on sandy drifts, sandstones, and sands. They tend to have a raw-humus layer, underlain by a bleached grey horizon from which virtually all free iron has been removed, and a B horizon which often shows a pan with iron and aluminium enrichment.

Finally, the important role of parent material can be seen in the development of rendzinas and various brown calcareous soils. The rendzinas are shallow soils formed over limestone (including chalk), with a dark-coloured organic horizon resting directly on weathered limestone. Secondary deposition of calcium carbonate may occur at depth. On harder limestones, red or brown calcareous soils may form. In these soils, which are mainly neutral or somewhat alkaline, the A horizon grades into a brown or reddish-brown calcareous B horizon.

---

* A form of soil organic matter formed in freely drained base-rich soils with good aeration. Plant growth is good and provides much litter which is mixed into the soil by a rich soil fauna.

Fig. 100. Generalized soil types.

(*Source*: Wiegand 1990, fig. 11.)

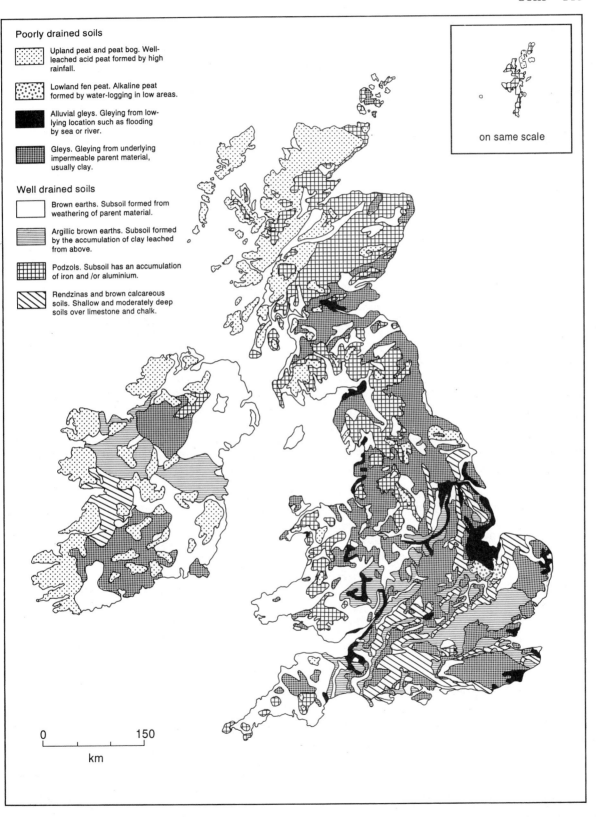

Poorly drained soils

Upland peat and peat bog. Well-leached acid peat formed by high rainfall.

Lowland fen peat. Alkaline peat formed by water-logging in low areas.

Alluvial gleys. Gleying from low-lying location such as flooding by sea or river.

Gleys. Gleying from underlying impermeable parent material, usually clay.

Well drained soils

Brown earths. Subsoil formed from weathering of parent material.

Argillic brown earths. Subsoil formed by the accumulation of clay leached from above.

Podzols. Subsoil has an accumulation of iron and /or aluminium.

Rendzinas and brown calcareous soils. Shallow and moderately deep soils over limestone and chalk.

on same scale

0   150
km

Soil erosion

The British Isles do not experience the very high rates of soil erosion that are encountered in those countries with more erosive rainfall regimes (e.g. the tropics). Rainfall in Britain tends to be much more evenly distributed in time and space so that a map of rainfall erosivity (Figure 101) shows an almost random pattern. There is a suggestion of higher energy towards the eastern half of the country due to the frequency of convectional rain.

Many of our soils have good structures and contain sufficient humus to give them resistance to attack by wind and water. Forests, sod-covered grasslands, and cropland offer a reasonable degree of protection. The soils most at risk are the sandy and sandy loam soils on slopes greater than 3° and where the land is under arable cultivation. In recent years it has become apparent that in these areas soil erosion is a significant problem. This is partly because miscellaneous land-use changes have rendered soils more prone to attack: the ploughing-up of ancient pastures, the removal of hedgerows, the increased adoption of autumn-sown cereals, the ploughing and drainage of peaty soils in both lowland (e.g. the Fens) and upland situations (e.g. newly afforested areas), and greater recreational pressures. Approximately 37% of the total arable area is at risk with a further 6% of non-arable land also vulnerable.

The areas which, because of their soil properties and climatic conditions, are potentially vulnerable to soil erosion by water, wind, or a combination of both, are shown in Figure 102 and include:

1. The Vale of York, parts of Nottinghamshire, and north Norfolk: sandy and coarse loamy soils are prone, when used for arable farming, to both water and wind erosion.

2. Much of the Midlands, Welsh Marches (particularly Shropshire and Herefordshire); a belt extending from Berkshire north-eastwards into Bedfordshire, Cambridgeshire, and Suffolk; parts of south Devon, south Somerset, Dorset, Isle of Wight, east Hampshire, and the North and South Downs: sandy, sandy loam, loamy, and silty soils are prone, when used for arable farming, to water erosion.

3. North Norfolk, east Suffolk, parts of Lincolnshire, Yorkshire, and Nottinghamshire, and areas of lowland peat soils in the Fens, western Lancashire and Somerset: sandy and sandy loam soils are prone, when used for arable farming, to wind erosion.

4. Extensive areas of the Pennines and the Welsh mountains, and small areas in the Lake District, Dartmoor, and Exmoor: blanket peats are vulnerable to water, especially gully erosion.

5. Areas of coastal sands: subject to wind erosion.

In these regions water erosion may not always be apparent to the eye because soil losses at a rate of perhaps 1 t ha y^{-1} tend to be insidious rather than dramatic. The depletion over time is due to raindrop splashing, overland flow, and shallow rilling. It is the rills which may occasionally be noted on bare soil after a storm. The controls are the rainfall energy and the angle and toughness of the slope: the soil properties and the use of the land.

Wind erosion tends to be more obvious because dust in the atmosphere or blowing across the surface tends to be very visible. Surprisingly it has received less attention unless a particular storm causes widespread damage or crop loss. The sensitive soils are again sandy loams and clay loams where organic content is low.

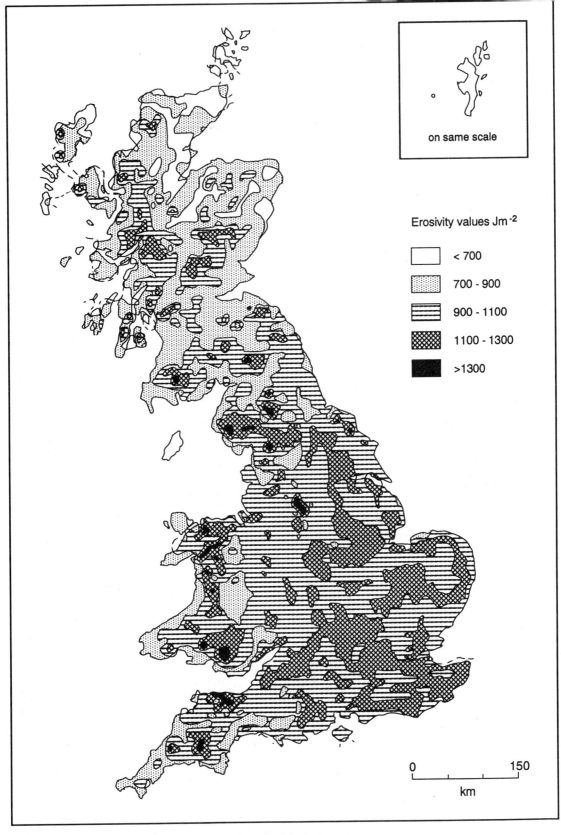

Erosivity values Jm^{-2}

	< 700
	700 - 900
	900 - 1100
	1100 - 1300
	>1300

on same scale

0   150

km

Fig. 101. Mean annual rainfall erosivity.

(*Source*: Morgan 1980.)

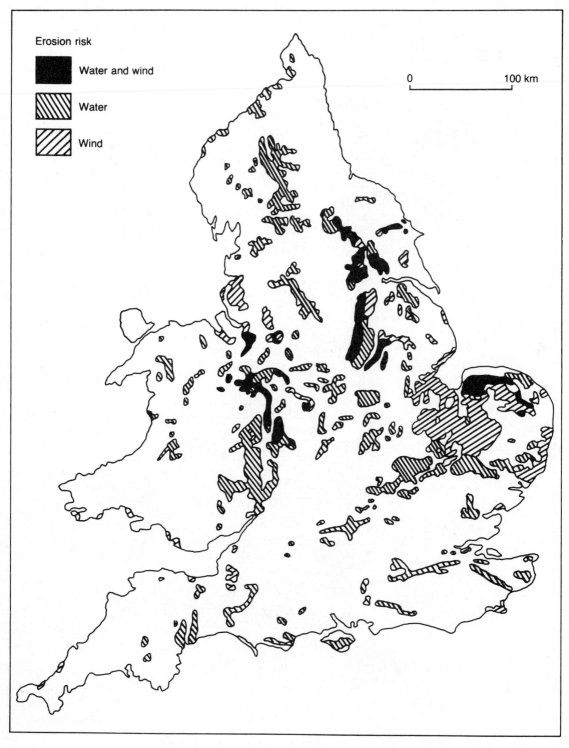

Fig. 102. Areas susceptible to agricultural soil erosion by wind and water.
(*Source*: Morgan 1986, fig. 4.17.)

Oceans

Tidal conditions

The tidal range is the difference in depth of water between high and low water. The largest range occurs during spring tides and the lowest during neaps. A useful measure of the range for geomorphological purposes is the mean spring tide. It should be remembered that these change by up to 30% in onshore gales or strong offshore winds. Unusual ranges occur if atmospheric pressure, wind, and tide interact to yield a tidal surge. Knowledge of surges is important for coastal-flood warnings.

The tidal range onshore is much larger than in open ocean water (Figure 103). The biggest tides occur in narrowing inlets such as the Bristol Channel (>9 m), and in estuaries and bays. These features are clearly seen along the Irish Sea and in the Mont-Saint-Michel–Channel Isles–Cherbourg area. The high tides on the east coast, which are unlike those elsewhere in the North Sea, are caused by the Coriolis force of the earth's rotation. This deflects any moving object to the right in the northern hemisphere. The tide moving south tends to pile up on the east coast. In such conditions a surge can develop and there can be severe risk of flooding. A resonance effect can also cause a double tide, such as at Southampton, where tides enter an inlet from two separate routes around the Isle of Wight.

The gravitational attraction in the sun and moon, and the rotation of the earth create weak tides in the major oceans. These are passed indirectly to the British seas where the geometrical configuration of the coast interacts with the tidal cycle. Here the weak Atlantic tidal stream is increased, first by shallowing and then, secondly, as it is forced through the narrow entrances of

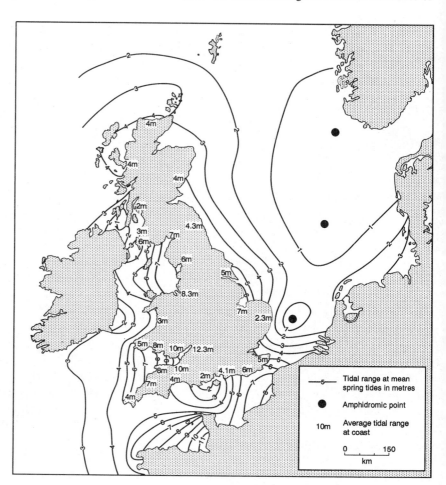

Fig. 103. Tidal range at mean spring tides.

(*Source*: NERC 1991; Pethick 1984.)

the Irish Sea, Dover Strait, and Pentland Firth. Headlands like Cherbourg, Portland Bill or the East Anglian coast cause tidal 'races', the highest tidal ranges being associated with the strongest streams. Very strong tides occur in the Galway Firth and the Bristol Channel which may have the second highest tide in the world after the Bay of Fundy. In the areas between Scotland and Ireland current speeds may reach 7.0 knots (3.6 m sec^{-1}). In the North Sea, however, speeds rarely exceed one knot (0.51 m sec^{-1}).

The tidal, wave, and configuration conditions of the British coast determine that there will be a *general* near-surface water movement pattern (Figure 104). The

actual movement of an object in the water will be affected by seasonal circulations and local gales but the overall direction will average out as a movement shown by the arrows on the map. The rate approximates to 1.5 nautical miles per day. The easiest way to interpret the map is to imagine you wish to send a 'bottle message'. A message sent from Grimsby would probably go to Holland or Germany!

Particularly important is the steady drift from the Atlantic past Ireland, Scotland, and Norway. This is the Gulf Stream or North Atlantic Drift, which has such a marked effect on the warmth of our northern waters and the mildness of our climate.

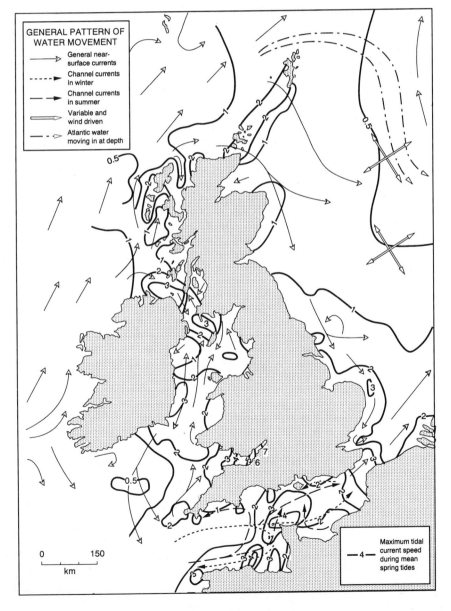

Fig. 104. Maximum tidal current speed during mean spring tides and general surface-water movement pattern.

(*Source*: NERC 1991.)

Sea-surface temperatures

The tidal stream and surface-water movement maps show how water from the North Atlantic Drift moves past the British Isles to Norway and to France. In winter this brings water that is warmer than the local waters around the west coast and two tongues of warmth enter the Irish Sea and the English Channel whilst a broad lobe extends down the North Sea.

The winter map is for February, the coldest month for the coast of Europe (Figure 105). When ice forms off the Dutch, German, and Danish coasts the Scottish coast can be 8 °C warmer and ice-free! The isotherms are strongly oriented N–S except, of course, along the English Channel. Both patterns show the very strong penetration of warm water. Meanwhile, the extreme cold of the north European coast is affected by river water as well as the continental climate air–sea interactions.

The chart for the monthly mean temperatures for August presents the opposite picture (Figure 106). Now the warmest waters occur along the Dutch, German, and Danish coasts, close to the north European river-water outlets, and the 'hot' European landmass. Western waters are now relatively cool. The isotherms are oriented E–W with a very regular distribution along the English Channel. Notably the European 'warmth' extends up the Norwegian coast and central North Sea. A strong tongue of cold water penetrates down the east coast giving the notorious 'bracing' beach conditions compensated only by good cold-water fishing! The Irish coast remains equable at just a few degrees warmer than winter. Only the enclosed bays of the west coast of England and Wales reach the comfortable temperatures of 15–18 °C.

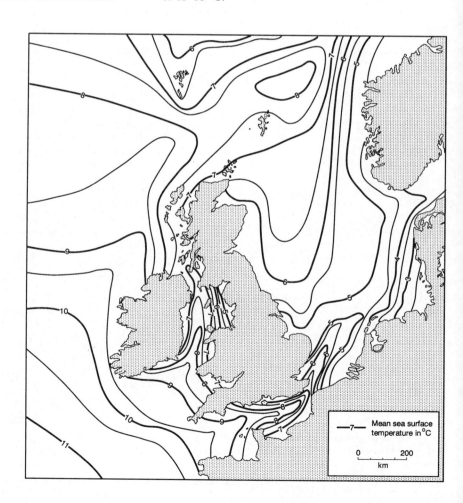

Fig. 105. Mean sea-surface temperature in winter.

(*Source*: NERC 1991.)

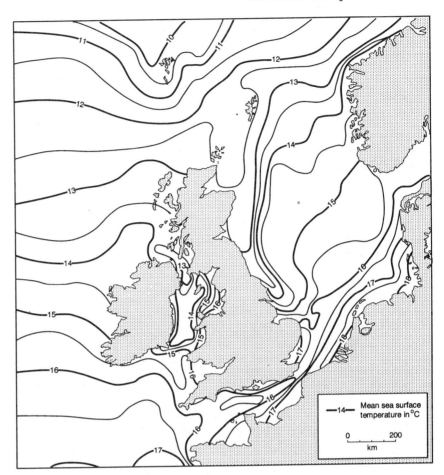

Fig. 106. Mean sea-surface temperature in summer.

(*Source*: NERC 1991.)

Waves

The way in which waves arrive at a coast is often described in terms of how often waves of a certain height are expected to occur. This is calculated statistically from past records and is expressed as a probability (e.g. on average, once in 5 years) or as a recurrence interval (the expected interval between events in years).

The coast of Britain is open to the Atlantic. The brunt of the wave energy is borne by western Ireland and the Scottish isles. Here the 50-year recurrence interval maximum wave height is 35 m with a period of 16 seconds. In the north, maximum 50-year wave heights of 30 m reach the Orkneys but in the North Sea, Irish Sea, and the English Channel these massive waves, and their associated energy, are quickly attenuated (Figure 107).

On the south coast, waves of 20 m occur in Lyme Bay, but by the time the Isle of Wight is reached the maximum 50-year wave reduces to 15 m, then to 10 m at Dover. The frequency with which they arrive reduces to 10 seconds. A similar pattern occurs towards the north Wales and Cumbrian coasts, and down the North Sea.

Attenuation in this pattern does not, however, take place along the east coast. Reference to Figure 107 shows that this coast suffers from tidal buildup due to the earth's rotation, experiences tidal surges as the tide moves south, and has a 500 km fetch if a storm strikes from the north. By themselves each of these figures give a rather innocuous pattern. Together they can combine to create the worst potential coast-flooding problem in Britain. In January 1953 sea-levels rose by more than 3 m above normal on the east coast and Thames estuary. 800 km^2 of land were flooded with a death of a total of 300 people. Today the community response has been to build the Thames Barrier at a cost of £450 million (1982). £300 million was spent on raising and strengthening the flood walls. London now has protection against a 7.2 m above mean sea-level flood.

Care should be taken in assessing the map. As the maximum waves move onshore their energy may be dissipated by contact with the sea floor. The waves convert to breaking waves and although these may rush high up a beach or cliff they are not fully effective. In a few locations, however, deep water comes close onshore. In these circumstances all the wave energy strikes the shore. Chesil Beach is one example. Here a 20 m offshore wave easily overtops the beach even though it is more than 15 m high. In fact this beach suffers in the 1 : 20 year wave and does not have to wait for the exceptional storm. A worthwhile point to make when considering coastal landforms therefore is whether the form reflects the maximum energy applied to it. On the west coast of Ireland, the Scilly Isles and Cornwall, or the Hebrides, the beaches and cliffs will be created by events of larger energy than those at, say, Brighton. This does not mean that the erosion or sediment transport rates will be higher, but only that the forms will be in equilibrium with more extreme conditions. Conversely, those coasts which are used to weaker energy applications are more 'surprised' when a very rare big event occurs.

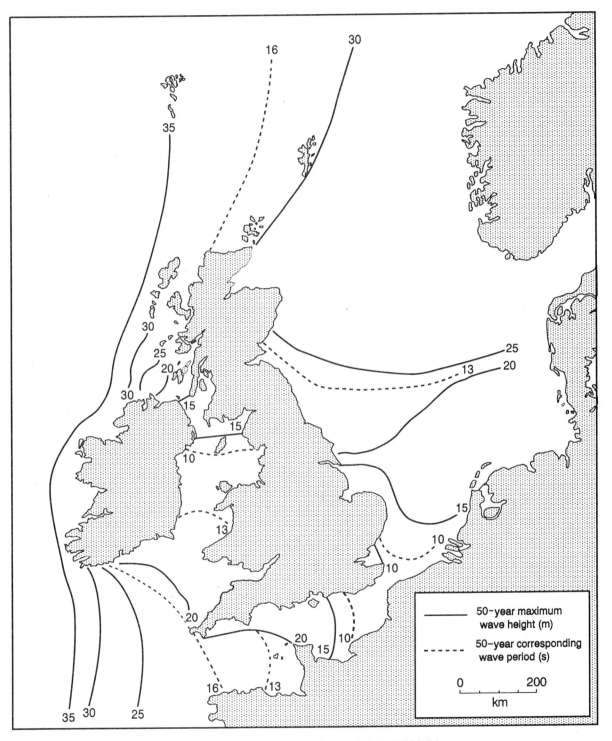

Fig. 107. Wave characteristics around the British Isles.
(*Source*: NERC 1991.)

Bottom deposits

The deposits on the floor of the sea are normally classified in terms of their texture (mud (silt-clay), sand, gravel, or mixtures) and rock outcrops. These are shown in Figure 108.

The origin of these deposits is very complex. Many were laid down on the floor of the North Sea, western Scotland, and Irish Sea by the ice sheets of the Quaternary period. These reached as far as north Devon and the Scilly Isles during the Anglian glaciation. When sea-level was perhaps 100 m below present, the sea floor south of these ice sheets was exposed. Meltwater from the ice, glacial lakes in the North Sea, and rivers from southern Britain, France, and Holland, spread gravelly debris as braided streams across this sea floor. In areas covered by the sea, floating ice carried drop stones and other debris which fell to the sea floor from the melting base. The Thames, Rhine, and Seine together with smaller rivers created deltas in proglacial lakes and in the Western Approaches. Beach deposits formed at the successive coastal positions and wind-blown deposits developed by winnowing of the fine debris.

As the sea rose it encroached on these deposits. The sea combed up sands and gravels with a succession of shorelines. Barrier beaches and bars were driven onshore and materials from the west were sorted along the channel toward the east. Thus Budleigh Salterton quartzites are found in beach materials and offshore banks as far east as Brighton.

Much coarse material was left behind as a basal layer and as beaches were submerged they were eroded to form the bars and banks of today. Much of the fine silts, clays, and sands were redistributed into hollows and backwater areas. Tidal amounts were very important in this process so that today many of the sediment cells operate with respect to relict stores of offshore material and pathways that were determined over a long period of tide, wave, and sea-level change. Failure to appreciate this has led to many misconceptions about the coast and many mistakes in coastal management.

Current deposits are also related to the maximum tidal-current velocities during mean spring tides and to long-term net sand-transport directions. Large dune forms, ripples, and other sediment structures show how dynamic this movement can be.

In the west there is more shell debris. In the North Sea the material is rock, fluvial, and glacial derived. This is reflected in the composition of the beaches and dunes (Figure 31). Mud is very common in the North Sea. This reflects the influence of fluvial sediment supply compared with the Atlantic coasts and the erodible rock types of lowland Britain and Europe. Approximately $9-10^6$ million tonnes is deposited in the North Sea each year. No wonder it is subsiding!

A final interesting point is that the continental slope is marked by numerous random furrows at depths of 140–500 m (Figure 108). Their size ranges from 20–100 m wide, 2–10 m deep, and 0.1–5.5 km long. They are regarded as grooves produced by icebergs calving from the front of the last Pleistocene ice sheets. Plough marks, raised rims, and crossing grooves support this idea and suggest that the icebergs were the size of the biggest found today.

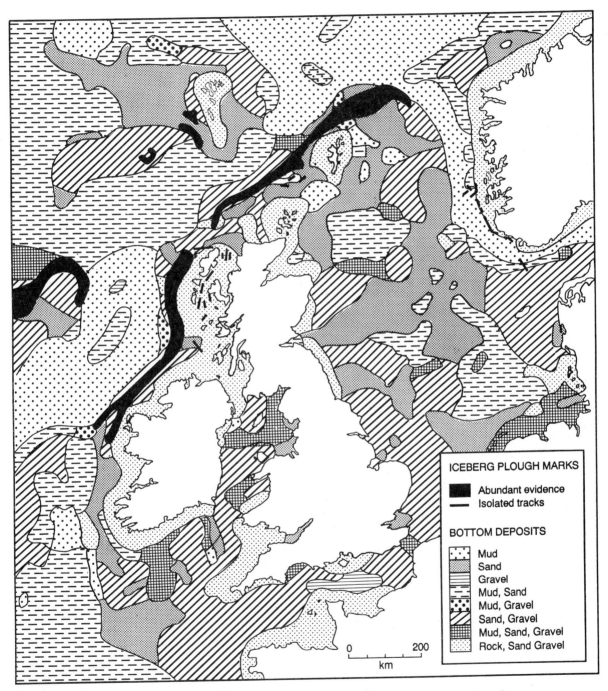

Fig. 108. Main types of bottom deposits and iceberg plough marks.
(*Source*: NERC 1991.)

ICEBERG PLOUGH MARKS

■ Abundant evidence
— Isolated tracks

BOTTOM DEPOSITS

Mud
Sand
Gravel
Mud, Sand
Mud, Gravel
Sand, Gravel
Mud, Sand, Gravel
Rock, Sand Gravel

0   200
km

The Human Impact

Interference with runoff

Human beings are unable to leave water alone and there is no such thing as a natural river in Great Britain! The most dramatic effect is the huge variety of ways in which stream channels can be modified. Figure 109 shows river channelization. These are sections which have been straightened, deepened, dredged, contained, or otherwise 'improved'. Channel improvements are largely designed to improve water flow and land drainage, or to prevent water loss. Bypass and diversion channels are produced to carry excess floodwater or for irrigation.

Figure 110a shows the streams affected by regulation. This is usually achieved by the construction of dams and reservoirs. In some upland areas as much as 20–70% of catchments are regulated in this way. Sometimes 1–9% of catchment areas are inundated with a 10–70% reduction in peak flows. Other consequences include: morphometric adjustments of channels, aggradation above and erosion below the structures, sediment retention leading to coastal erosion, changes to groundwater levels, and ecological effects.

Abstraction of water (Figure 110b) is another serious change to the natural fluvial system. This again leads to morphological and ecological changes, sometimes very serious in time of drought or sequences of abnormally dry years and low water-tables. Since 1986 many people have complained of the loss of their rivers due to too much abstraction from boreholes.

The construction of ponds, dells, pits, mires, and other holes in the ground including subsidence hollows caused by the abstraction of fluids, coal, or minerals, is also typical of the British landscape. There may be as many as 350 000 ponds in England and Wales alone and this is a much reduced figure from the maximum of over 800 000 in the nineteenth century. Figure 111 shows that the lowlands of Britain have by far the highest concentration. Norfolk and Suffolk have over 6 ponds per km^2, and Cheshire, which also has subsidence ponds, has over 20 ponds per km^2. Other local concentrations are in mining areas such as Cornwall, Durham, and the coalfield areas. The effect on the water system is unknown but certainly the local effects on the ecosystem are substantial.

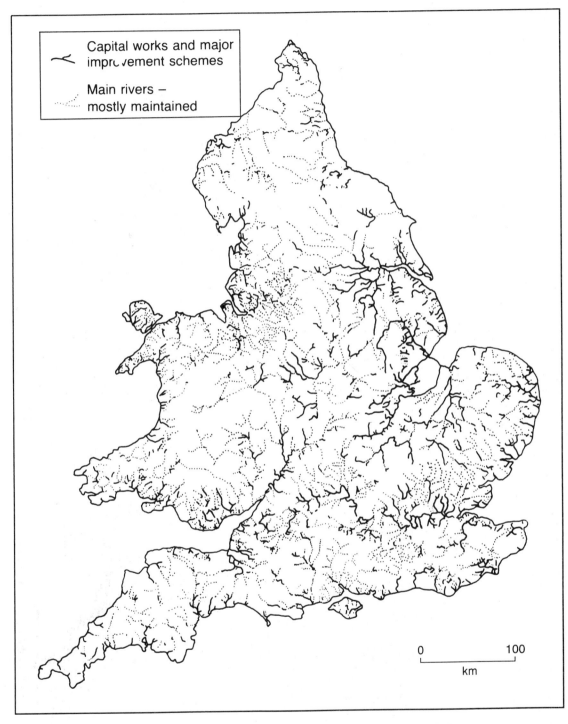

Fig. 109. Rivers that were channelized between 1930 and 1980.

(*Source*: Brookes *et al.* 1983.)

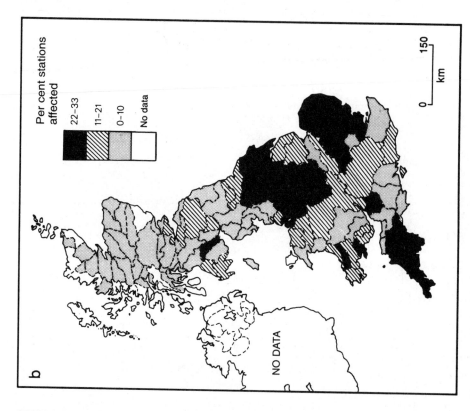

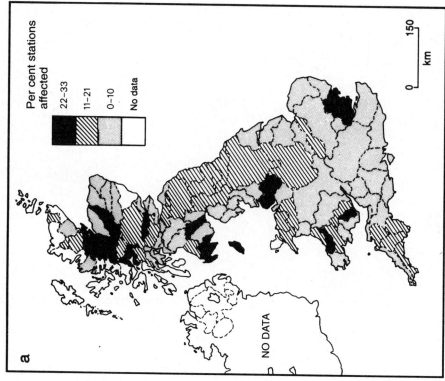

Fig. 110. Patterns of interference with streams: (*a*) Regulation; (*b*) Abstraction.

(*Source:* Lewin *et al.* 1981, fig. 1.5.)

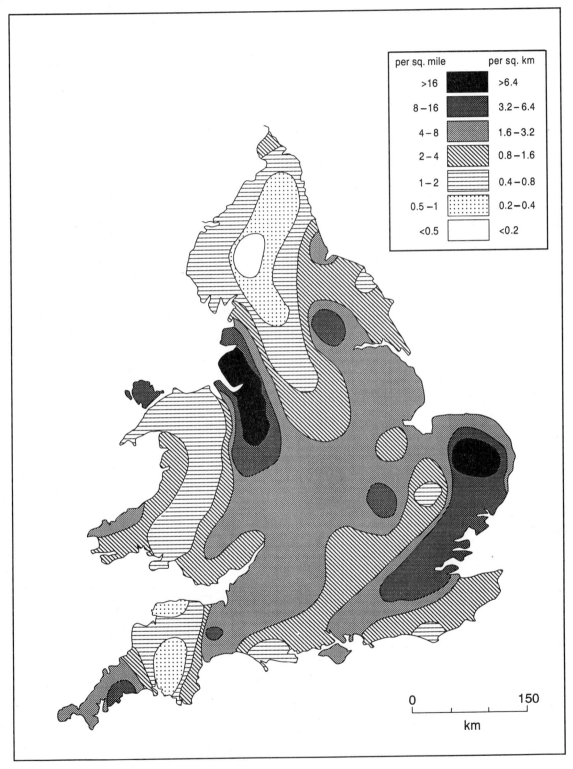

Fig. 111. Distribution of ponds in the 1920s.

(*Source*: Rackham 1986, fig. 16.1.)

Ozone

Ozone is a gas that is extremely reactive and which is potentially toxic to humans and plants. It occurs naturally in the atmosphere, but higher than natural levels can be produced by the action of sunlight on the nitrogen oxides and hydrocarbons that are emitted in exhaust gases produced by the burning of fossil fuels (e.g. in cars). The effects of ozone on humans include respiratory complaints, eye irritation, and headaches. It may be toxic to many species of conifer, herbaceous plants, and crops at concentrations not far above the natural background level.

In contrast with sulphur and nitrogen, ozone does not persist or accumulate in ecosystems. As a consequence, neither the geographical pattern of dispersal nor its effects can be determined by analysing residues on soil, vegetation, water, or sediments. The environmental concern is, therefore, not with the buildup of residues in ecosystems but with the concentrations of ozone that build-up in the air under appropriate conditions.

The concentrations of ozone depend on the availability of sunshine, the presence of suitable gases for reaction in sunlight, and the strength and trajectories of dispersal mechanisms (especially wind). The first of these controls—sunshine—is so variable in Britain that ozone tends to occur in episodes and to vary in its concentrations and patterns from day to day, season to season, and year to year. Figure 112a is for 1987, a dull and relatively sunless year, and it shows the concentrations of ozone expressed as the number of hours with a mean concentration of over 60 parts per US billion. The gradient in concentration outwards from London is due partly to the time taken for ozone to form and partly to the fact that some of the precursor gases over the metropolis also destroy ozone. Especially high values occur on the south-east coast of England, partly due to the import of ozone from the continent. Figure 112b is for 1990—a sunny year—and uses the same units. The concentrations are very much higher for the whole country, with the highest values in the south of Wales, the West Country, and south-east England.

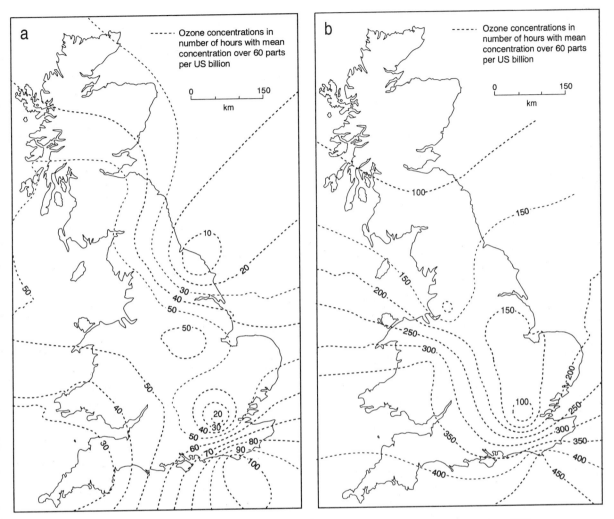

Fig. 112. (*a*) Ozone concentrations 1987; (*b*) Ozone concentrations 1990.

(*Source*: Davison and Barnes 1992, figs 6.6*a*, and 6.6*b*.)

Acid rain: The wet deposition of pollutants

One of the most celebrated environmental issues of the last few decades has been the question of the extent to which the rainfall of the British Isles have been rendered acidic by various human activities, including the burning of fossil fuels (Figure 113). The sulphates released by the burning of coal and natural gas in power-stations and elsewhere, cause acidity because they lead to the production of weak sulphuric acid in rain. Nitrates released from car exhausts cause the development of weak nitric acid; ammonium produced by agriculture (livestock waste) is another significant pollutant. Acid rains are not the only source for raindrops can scavenge from acid concentrations in cloud-water over high ground. Acid rain droplets are concentrated near the cloud base and it is now thought that this has a severe effect on plants.

Normal rainfall tends to have an acidity value (pH) of around 5.6 to 5.7, so regions with rain acidity less than this value can be classified as acidified. Recent work, which has made allowance for scavenged cloud-water, shows very large increases over hills (Table 6). The areas where the greatest increases occur are those where orographic enhancement of deposition is prominent and is related to the elevation. In the west of Britain, for example, orographic enhancement of precipitation is threefold over the coastal rate. Acid deposition reflects this but if allowance is made for rain seeding from concentrated clouds the acid deposition is increased fivefold. Areas with lower rainfalls like the Cairngorms or Yorkshire Moors, however, had lower concentrations.

The consequences of acidification are legion and include declines in fish stocks, the release of toxic aluminium into drainage waters, and the accelerated weathering of building stone. Not all parts of the British Isles will show the same response to acid rain as others. For example, areas where the soils are naturally alkaline (e.g. in limestone areas) are less prone to severe acidification than areas where there is less alkaline

material to buffer the effects of acid rain inputs. In general, areas underlain by old igneous and metamorphic rocks with large falls of rain are most likely to suffer. Figure 113 shows the areas which are thought to be rendered susceptible because of their soils and solid geology. The upland, western parts of the country are dominant in this pattern, but areas underlain by sandstones (like parts of the Weald) or sandy gravels (like the New Forest) may also be susceptible.

Figure 113 also shows those areas where water analyses demonstrate acidic surface waters. As we have already seen, normal rainfall tends to have an acidity (pH) value of around 5.6 to 5.7, so regions with freshwater acidity less than this value can be classified as acidified. There is a reasonable correspondence with the susceptibility pattern.

One of the most successful ways to identify trends in acidity is to take cores from lake floors and to look at the diatom flora from layers of different ages. These particular algae have a distribution that is strongly controlled by the acidity of the water in which they live. Historical analysis of diatom-rich lake sediments indicates that prior to the industrial revolution very few lakes had acidic waters. In general the trend to lower pH values began around or after 1850 and generally involved a decline of between 0.5 to 1.5 pH units from pre-1850 levels. An acceleration of pH values occurred between 1930 and 1970. Because of a recent improvement in sulphate emissions from power-stations, and various other sources, some lakes may be starting to recover.

It should also be noted that there is strong evidence from upland areas of Scotland, England, and Wales that streams draining coniferous plantations are more acid than those draining grassland and moorland. That is why the new work on turbulent cloud feeding is so important. Conifers enhance turbulence and are capable of taking considerable moisture from the cloud base just where the concentrations are highest. Drip and stem flow then supply the enhanced acid water.

Table 6. Acid rain in highland areas

	Snowdonia	Lake District	York Moors	West Highlands	Cairngorm
Precipitation (mm)	2900	2940	900	3220	1380
Acid deposition (mg m^{-2})	93	132	82	129	69
Sulphate (g m^{-2} S)	3	3.6	1.5	2.8	1.1
Nitrate (g m^{-2} N)	1.15	1.43	0.72	1.15	0.58
Ammonium (g m^{-2} N)	1.56	2.62	0.85	1.14	0.44
% increase in deposition	73	76	41	73	50

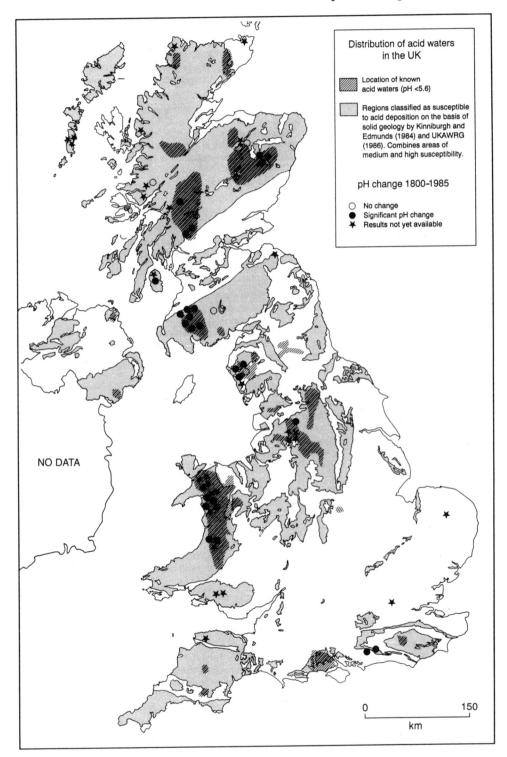

Fig. 113. Distribution of acid waters.

(*Sources*: UK Acid Waters Review Group, 2nd. Ann. Report 1988, Acidity in UK Fresh Waters; Battarbee *et al.* 1986, Map of core sites showing pH change, 1800–1985.))

Sulphur deposition

The pollution of air by the emissions of sulphur consequent upon the burning of fossil fuels has been a long-
term environmental concern in the British Isles. The
presence of sulphur in the air reduces air quality and
thereby affects organisms, including humans. It also
contributes markedly to the acidification of precipitation, thereby setting in train a series of environmental
consequences that range from forest decline, to poison-
ing of lakes, and to building-stone decay.

The emissions of sulphur dioxide in Britain are far
from uniform in their distribution. As one might anticipate, the highest levels occur in large industrial areas
and in proximity to major non-nuclear power-stations.
High values occur in west Pembrokeshire (Dyfed) and
near Southampton associated with the oil refineries and
oil burning power-stations of Milford Haven and
Fawley respectively. More generally, high levels are
emitted from the London region, south Wales, the
Midlands, north-east England, and the Central
Lowlands of Scotland. Figure 114 records the annual
rate of dry deposition of sulphur in grams per square
metre. A broadly similar pattern to that of emissions is
evident, and notably low values of deposition occur in
western areas. The belt of highest deposition stretches
across England from the Dee to the Humber.

Maps for the annual deposition of nitrate $(g\,m^{-2}N)$
for 1986–8 and ammonium $(g\,m^{-2}N)$ 1986–8 are also
shown for comparison (Figures 115, 116). Mean SO_2
concentrations (above $30\,\mu\,g\,m^{-3}$) occur over all of central England, with levels below this being restricted to
south-west and northern England, nearly all Wales, and
all Scotland except the central valley. This pattern is
indicated by the map of the distribution of lichens
(Figure 117). These organisms are very good indicators
of levels of atmospheric pollution. There are some quite
extensive areas, notably in a belt between London and
Lancashire, where lichens are sparse or absent.

In recent years, because of changes in industrial tech-
nology and changes in pollution regulation, sulphate
emissions have shown a welcome decline in most areas
and many of these patterns, based on data for the 1970s
and 1980s, may now show some change.

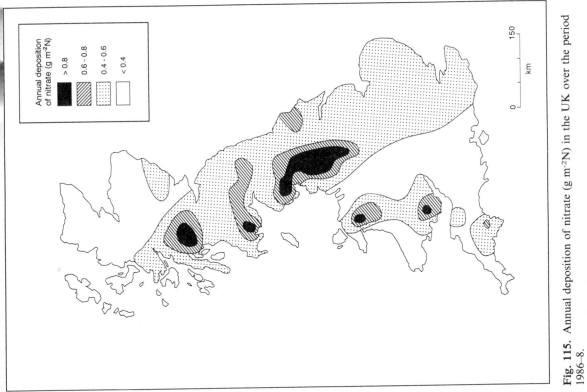

Fig. 115. Annual deposition of nitrate (g m⁻²N) in the UK over the period 1986–8.

(From Dore *et al.* 1992, fig. 5.)

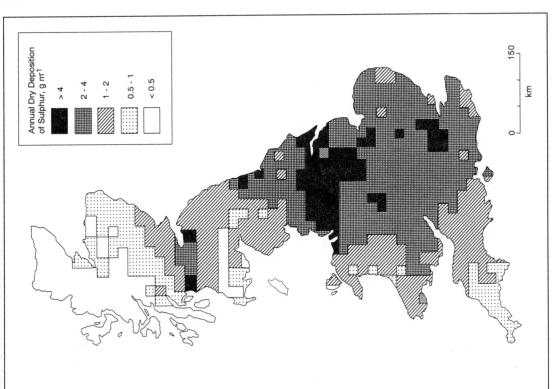

Fig. 114. Annual dry deposition of sulphur, g m⁻².

(*Source:* Barrett *et al.* 1983.)

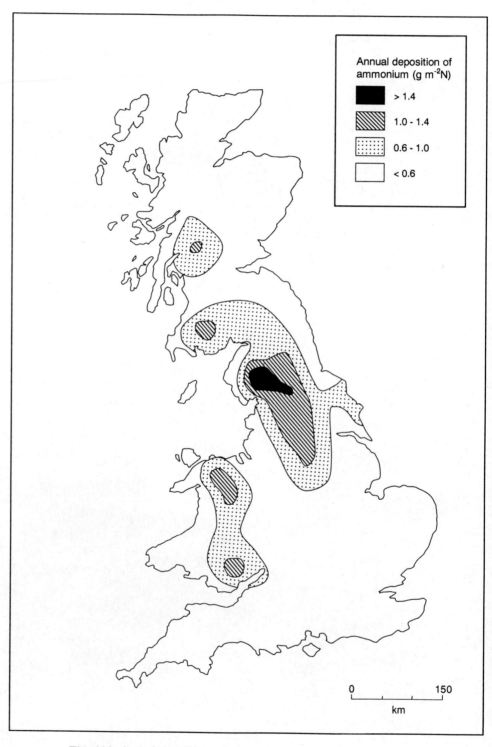

Fig. 116. Annual deposition of ammonium (g m⁻²N) in the UK averaged over the period 1986–8.

(From Dore *et al.* 1992, fig. 6.)

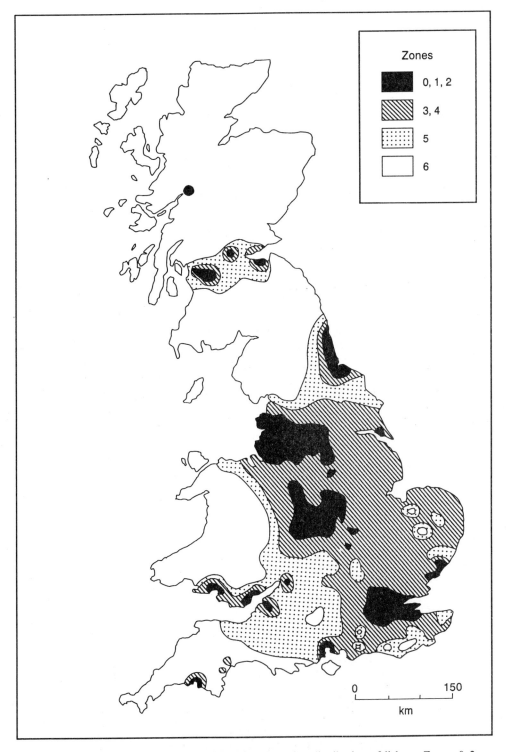

Fig. 117. Zone map of Great Britain showing distribution of lichens. Zones: 0–2, no lichens or very sparse; 3–4, leafy lichens; 5, some shrubby lichens; 6, lichen *Usnea* and other species indicating very low SO₂ levels.

(*Source*: Gilbert 1974.)

Dutch elm disease

Elm disease is caused by a microscopic fungus, the *Ceratocystis ulmi*, which affects the outer wood. It blocks the water-conducting vessels and interferes with the hormones. The tree wilts and whole limbs can bend over. The fungus is transported by bark-boring beetles or collected by roots.

The latest episode of the disease flared up in the late 1960s. It began in Tewkesbury in 1965, followed by Bristol, Southampton, London, Plymouth, and Ipswich. The first inland outbreaks were in north-east Hampshire, Breckland, and Buckinghamshire. The disease radiated out and travelled at perhaps 12 km per year. The outbreak centres joined up and began to destroy almost every large elm in the south. Some areas near Ipswich, Plymouth, and Essex curiously survived and parts of Kent and Cornwall were only partially affected. By 1972 the wave of disease had reached a line drawn from Liverpool to the Wash. By 1974 Wales and northern England had succumbed, and by 1977 most of Scotland.

The most susceptible elm is the English elm which occurs naturally throughout the area of disease origin. Today, although suckers survive, most large trees between Manchester, Chichester, and Exeter are dead and have been cut down. Where there are survivals this is mainly due to the prompt action of local authorities who burnt infected trees. Wych elms and Irish varieties are unevenly affected. The Huntingdon elm and those in Cambridge and Essex seem to be more resistant.

Figure 118 shows the effects only of the latest most virulent epidemic. Since the disease was first noted in France in 1918, however, it has occurred several times. It acquired its name after a severe epidemic in Holland. It first occurred in Britain in 1927, peaking in 1936. It then declined leaving many trees intact and perhaps 80% survived or recovered. The mysterious decline of the disease, which has occurred in each epidemic is, however, not fully understood.

Finally, it should be noted that there are historical records of earlier destructions of elms. In these cases the disease has not been identified but circumstantial evidence suggests that the infection may have been around for a long time. In 1818 trees died in Cambridge and London and by the 1830s and 1840s in France, Belgium, and the Netherlands, with a decline towards 1864. There are similar reports for 1780 in Oxford, and the cause of the Elm Decline in the Neolithic (5000 years ago) may have been disease rather than early forest clearance.

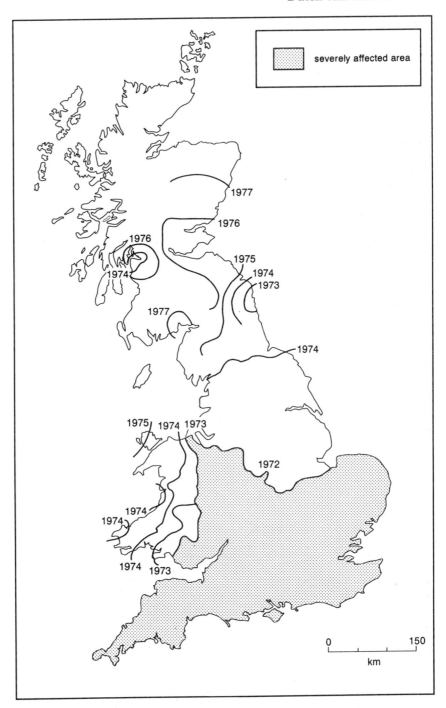

Fig. 118. The spread of Dutch elm disease.

(*Source*: Perry 1981, fig. 9.2.)

Oil pollution

The seas around the British Isles are becoming increasingly prone to contamination from oil because production, transport, and use of this commodity is expanding. The sources of this pollution include tanker collisions; the explosion of individual ships because of unsafe gas levels; tanker wrecks because of navigational error, mechanical failure, or storms; seepage from offshore installations; major platform accidents and the flushing of tanker holds and bilge pumping. There are some natural seepages but the main causes are human actions.

In the past the discharge of ballast water from empty tankers was a primary cause but this has been ameliorated by technical loading improvements. A general awareness and a tightening of regulations has done much to improve matters. Despite this there are major accidents with a consequent danger to bird and marine life.

It is not generally appreciated just how many incidents there are. Accidents like the Torrey Canyon of 1967 are well publicized. Figure 119 shows the incidents for just one year, 1977. London had 82 spills followed by the Medway (30), Clyde and Forth (17), and Manchester (14). The south coast had 56 spills within one mile, 10 off-shore, and well over 100 at sea.

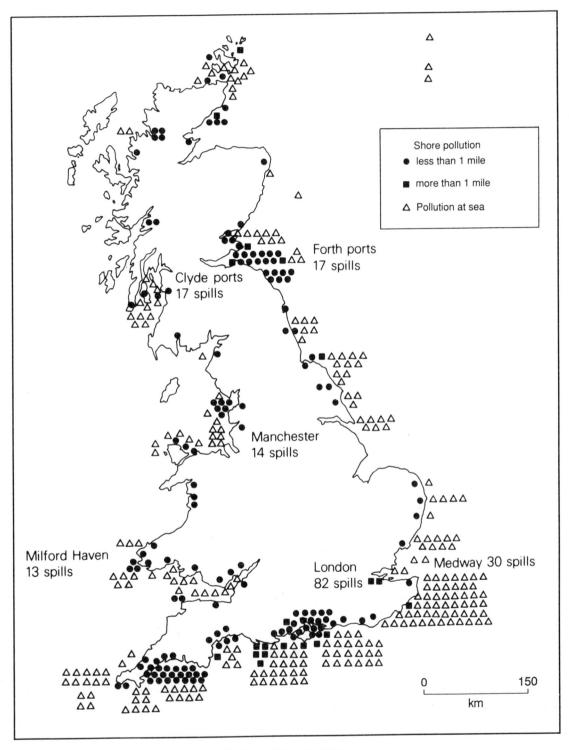

Fig. 119. Oil pollution incidents in 1977.

(After Perry 1981.)

Dumping at sea

One of the most extraordinary acts of pollution perpetrated by the British people is the dumping of waste at sea. This is governed by two international conventions, the Oslo Convention for the Prevention of Marine Pollution, which gives general guidelines, and the London Convention on the Dumping of Wastes, which applies world-wide. Locally, the Dumping at Sea Act 1974 provides legislation. It is an offence for any vehicle, ship, aircraft, hovercraft, or marine structure to dump at sea without a licence. The approximate location of dumps is shown in Figure 120 for dredge spoil, and Figure 121 for sewage sludge, and industrial, colliery, and spoil waste. Sewage-sludge licences range from 200 to 71 000 thousand tonnes per annum, industrial waste from 200–1000 thousand tonnes per annum, colliery waste 500–1500 tonnes per annum, and spoil from 50 000 to 10 million wet tonnes per annum.

The UK drops substantially more sewage sludge than other countries bordering the North Sea largely, it is claimed, because we treat more and have more access to rapid dilution tidal streams. This really means we can send it elsewhere more quickly. For example, 5 billion wet tonnes per annum are deposited in the Thames estuary. Quality ranges from primary to biological and digested sludge, usually of domestic origin but occasionally with industrial contaminants.

Industrial waste is normally dumped from vessels of 150–2000 tonnes capacity. The waste is not containerized. This dumping only takes place in deep water, 250 km from land and in water 2000 m deep. Most industrial wastes have low toxicity. Colliery waste or mine stone also includes fly ash from power-stations. It is often tipped on the foreshore in the hope that it will finish up below the high-water mark but some is barged offshore. The total is 2 million tonnes per year. Dredge spoils include spoil from maintenance dredging of channels and harbours, new docks, and other construction sites. The material ranges from silts, to sands and gravels, and some rock or compact clay. Approximately 70 sites are used for wet-spoil dumping. For example 18 million wet tonnes are deposited in the Bristol Channel.

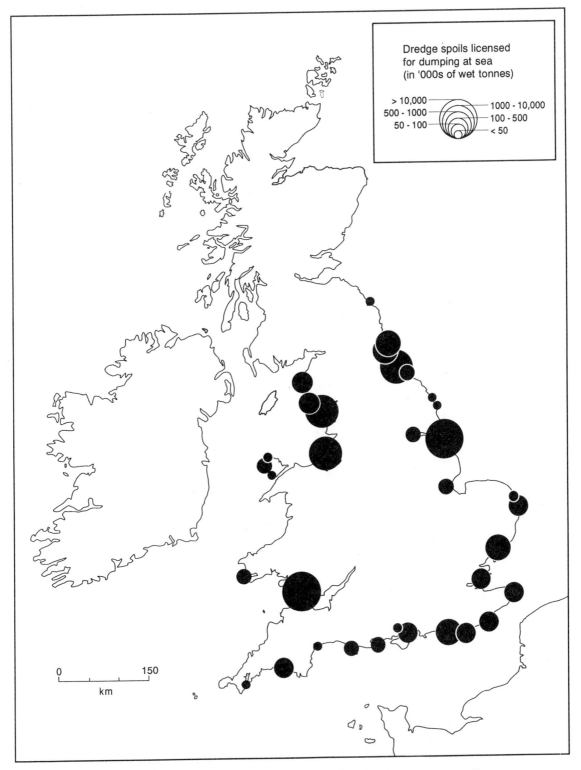

Fig. 120. Location of sites licensed for the dumping of dredge spoils at sea.
(*Source*: NERC 1991.)

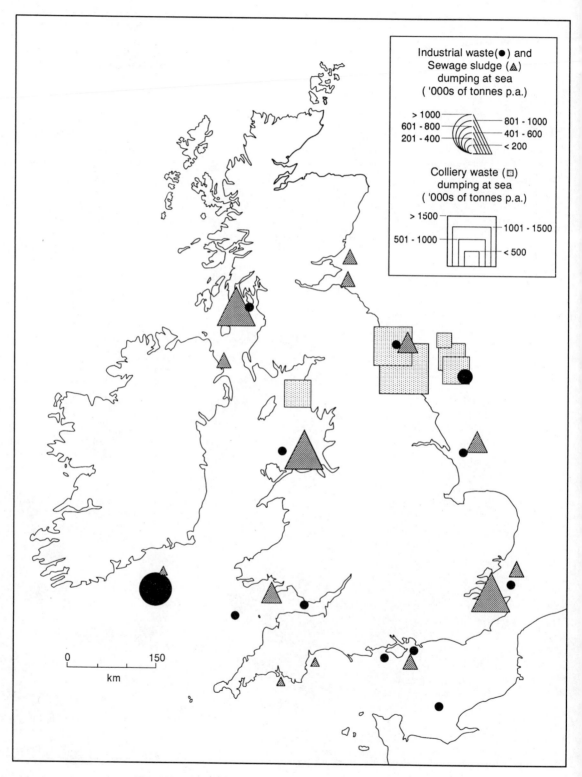

Fig. 121. Sites licensed for the dumping of sewage sludge, industrial, and colliery waste at sea.

(*Source*: NERC 1991.)

Acknowledgements

We would like to acknowledge the following sources of material:

Adam, P. (1990) *Saltmarsh ecology* (Cambridge University Press).

Allen, J. and Bird, R. A. (1977), *The prospects for the generation of electricity from wind in the United Kingdom* (HMSO: London).

Anderton, R., Bridges, P. H., Leeder, M. R., and Sellwood, B. W. (1979), *A dynamic stratigraphy of the British Isles* (Allen and Unwin: London).

Ballantyne, C. K. (1987), 'The present day periglaciation of upland Britain', in J. Boardman (ed.), *Periglacial processes and landforms in Britain and Ireland* (Cambridge University Press), 113–26.

Barrett, C. P. and Irwin, J. G. (1983) (eds.), *Acid deposition in the United Kingdom* (Warren Spring Laboratory: Stevenage, Herts.).

Battarbee, R. W. and collaborators (1986), *Lake acidification in the United Kingdom 1800–1986* (Ensis: London).

Bennett, K. D. (1989), 'A provisional map of forest types for the British Isles, 5000 years ago', *Journal of Quaternary Science*, 4: 141–4.

Betton, C., Webb, B. W., and Walling, D. E. (1991), 'Recent trends in NO3-N concentration and levels in British Rivers', *IAHS Publication*, 203: 169–80.

Birks, H. J. B. (1990), 'Changes in vegetation and climate during the Holocene of Europe', in M. M. Boer and R. S. de Groot (eds.), *Landscape: Ecological impact of climatic change* (IOS Press: Amsterdam), 133–58.

Boulton, G. S., Cox, F., Hart, J., Clayton, K. M., and Kenning, M. J. (1977), 'A British ice-sheet model and patterns of glacial erosion and deposition in Britain', in F. W. Shotton (ed.), *British Quaternary studies: Recent advances* (Oxford), 231–46.

Brookes, A., Gregory, K. J., and Dawson, F. H. (1983), 'An assessment of river channelization in England and Wales', *Science of the Total Environment*, 27: 97–112.

Brugge, R. (1991), 'The record-breaking heatwave of 1–4 August 1990 over England and Wales', *Weather*, 46: 2–10.

Brunsden, D., Gardner, R. A. M., Goudie, A. S., and Jones, D. K. C. (1988), *Landshapes* (David and Charles: Newton Abbot).

Burt, S. O. and Mansfield, D. A. (1988), 'The Great Storm of 15–16 October 1987', *Weather*, 43: 90–108.

Carter, R. W. G. (1988), *Coastal environments: An introduction to the physical, ecological, and cultural systems of coastlines* (Academic Press: London).

—— (1989), 'Rising sea-level', *Geology Today*, 5: 63–7.

—— (1992), 'Coastal conservation', in Institute of Civil Engineers, *Coastal management 92* (Thomas Telford: London).

Collier, C. G. (1990), 'Assessing and forecasting extreme rainfall in the United Kingdom', *Weather*, 45: 103–12, table 1.

Chandler, T. J. and Gregory, S. (1976), *The climate of the British Isles* (Longman: London).

Davison, A. and Barnes, J. (1992), 'Patterns of air pollution: Critical loads and abatement strategies', in M. Newson (ed.), *Managing the human impact on the natural environment* (Belhaven Press: London).

Doody, P. (1985) (ed.), *Sand dunes and their management* (Nature Conservancy Council: Peterborough).

Dore, A. J., Choularton, T. W., and Fowler, D. (1992), 'An improved wet deposition map of the United Kingdom incorporating the seeder–feeder effect over mountainous terrain', *Atmospheric Environment*, 26A: 1375–81.

Dunning, F. W., Mercer, I. R., Owen, M. P., Roberts, R. H., and Lambert, J. L. M. (1978), *Britain before men* (HMSO: London).

Elsom, D. M. and Meaden, G. T. (1984), 'Spatial and temporal distribution of tornadoes in the United Kingdom 1960–1982', *Weather*, 39: 317–23.

Gilbert, O. L. (1974), 'Air pollution survey by school children', *Environmental Pollution*, 6: 175–80.

Goudie, A. S. (1990), *The landforms of England and Wales* (Blackwell: Oxford).

Gregory, S. (1964), 'Climate', in J. W. Watson and J. B. Sissons (eds.), *The British Isles: A systematic geography* (Nelson: London).

—— (1955), 'Some aspects of the variability of annual rainfall over the British Isles for the standard period 1901–30', *Quarterly Journal of the Royal Meteorological Society*, 81: 257–62.

Hydrological data, United Kingdom 1988 Yearbook (1989) (Institute of Hydrology: Wallingford).

Jackson, M. C. (1977a), 'The occurrence of falling snow over the UK', *Meteorological Magazine*, 106: 26–38.

—— (1977b) 'Evaluating the probability of heavy rain', *Meteorological Magazine*, 106: 185–92.

Jones, B. and Mattingley, J. (1990), *An atlas of Roman Britain.* (Basil Blackwell: Oxford).

Jones, D. K. C. (1985), 'Shaping the land: The geomorphological background', in S. R. J. Woodell (ed.), *The English landscape, past, present, and future* (Oxford University Press), 4–47.

Lacy, R. F. (1977), *Climate and building in Britain* (HMSO: London).

Lewin, J. (1981) (ed.), *British rivers* (Allen and Unwin: London).

Lewis, G. (1991), in A. Perry and L. Symons, 'Driving Forces', *Geographical Magazine,* 13: 41.

Lovell, J. P. B. (1986), 'Cenozoic', in K. W. Glennie (ed.), *Introduction to the petroleum geology of the North Sea* (2nd edn.) (Oxford: Blackwell Scientific Publications), 179–96.

Lowe, J. J. and Walker, M. J. C. (1984), *Reconstructing Quaternary environments* (Longman: London).

Marsh, T. and Monkhouse, R. (1990), 'Hydrological aspects of the development and rapid decay of the 1989 drought', *Weather,* 45: 290–9.

Mayes, J. C. (1991), 'Recent trends in summer rainfall in the United Kingdom', *Weather,* 46: 190–6.

Morgan, R. P. C. (1980), 'Soil erosion and conservation in Britain', *Progress in Physical Geography,* 4: 24–47.

—— (1986), *Soil erosion and conservation* (Longman: London).

Musk, L. F. (1991), 'The fog hazard', in A. H. Perry and L. J. Symons (eds.), *Highway Meteorology* (Spon: London), 91–130.

NERC (1991), *United Kingdom digital marine atlas* (NERC: Swindon).

Newson, M. D. (1978), 'Drainage-basin characteristics, their selection, derivation and analysis for a flood study of the British Isles', *Earth Surface Processes,* 3: 277–93.

Perry, A. H. (1981), *Environmental hazards in the British Isles* (Allen and Unwin: London).

Pethick, J. (1984), *An introduction to coastal geomorphology* (Edward Arnold: London).

Rackham, O. (1986), *The history of the countryside: The full fascinating story of Britain's landscape* (J. M. Dent: London).

Rodda, J. C. (1970), 'Rainfall excesses in the United Kingdom', *Trans. Institute of British Geographers,* 49: 49–60.

Rose, J., Boardman, J., Kemp, R. A., and Whiteman, C. A. (1985), 'Palaeosols and the interpretation of British Quaternary stratigraphy', in K. S. Richards, R. R. Arnett, and S. Ellis (eds.), *Geomorphology and Soils* (Allen and Unwin: London), 348–75.

Russell, D. J. and Eyles, N. (1985), 'Geotechnical characteristics of weathering profiles in British overconsolidated clays (Carboniferous to Pleistocene)', in Richards, Arnett, and Ellis (eds.), *Geomorphology and soils,* 417–36.

Shennan, I. (1989), 'Holocene crustal movements and sea-level changes in Great Britain', *Journal of Quaternary Science,* 4: 77–89.

UK Acid Waters Review Group (1988), 2nd Annual Report.

Walling, D. E. and Webb, B. W. (1981), 'Water quality', in J. Lewin (ed.), *British rivers* (Allen and Unwin: London), 126–69.

—— —— (1986), 'Solutes in river systems', in S. T. Trudgill (ed.), *Solute processes* (John Wiley: Chichester), 251–327.

Ward, R. C. (1981), 'River systems and river regimes', in Lewin (ed.), *British rivers.* 1–33, table 1.1.

Warren, W. P. (1987), 'Periglacial periods in Ireland', in Boardman (ed.) *Periglacial processes and landforms in Britain and Ireland.* 101–12.

Watson, J. W. and Sissons, J. B. (1964), *The British Isles: A systematic study* (Nelson: London).

Whittle, I. R. (1989), 'Landslide risk from sea-level rise in the UK', in J. C. Doornkamp (ed.), *Greenhouse effect and rising UK sea-levels* (M1 Press: Nottingham), 85–93.

Wiegand, P. (1990) (ed.), *The new Oxford school atlas* (Oxford University Press).